Asia-Pacific Fishing Livelihoods

Michael Fabinyi • Kate Barclay

Asia-Pacific Fishing Livelihoods

palgrave
macmillan

Michael Fabinyi
Faculty of Arts and Social Sciences
University of Technology Sydney
Sydney, NSW, Australia

Kate Barclay
Faculty of Arts and Social Sciences
University of Technology Sydney
Sydney, NSW, Australia

ISBN 978-3-030-79590-0   ISBN 978-3-030-79591-7   (eBook)
https://doi.org/10.1007/978-3-030-79591-7

Cover pattern © Melisa Hasan

This Palgrave Macmillan imprint is published by the registered company Springer Nature Switzerland AG.
The registered company address is: Gewerbestrasse 11, 6330 Cham, Switzerland

Acknowledgements

The research on which this book was based on was supported by multiple funded projects, including the Australian Research Council (ARC) (DP140101055, DP180100965), the David and Lucile Packard Foundation (2014-40093, 2017-65792), the Fisheries Research and Development Corporation (2014-301, 2017-092) and a Society in Science Branco Weiss Fellowship (Fabinyi). We also acknowledge the support of the Faculty of Arts and Social Sciences at the University of Technology Sydney, the Nippon Foundation Ocean Nexus Center, the ARC Centre of Excellence for Coral Reef Studies at James Cook University and the Crawford School of Public Policy at The Australian National University.

We thank our colleagues with whom we have collaborated in these projects and who helped to contribute to the ideas and arguments in the book, in particular Dedi Adhuri, Kirsten Abernethy, Eddie Allison, Wolfram Dressler, Hampus Eriksson, Simon Foale, Jeff Kinch, Nick McClean, Michael Pido, Andrew Song, Dirk Steenbergen, Reuben Sulu and Michelle Voyer.

CONTENTS

About the Authors

Michael Fabinyi is an associate professor in the Faculty of Arts and Social Sciences at the University of Technology Sydney (UTS). His research interests are focused on the social and political aspects of marine resource management and use, including coastal livelihoods, fisheries governance and fisheries trade.

Kate Barclay is a professor in the Faculty of Arts and Social Sciences at UTS. Her research interests are the governance of marine areas and resources, including social and economic seafood value chains, social inclusion in fisheries and the wellbeing of people in coastal communities.

ABBREVIATIONS

EBFM	Ecosystem-based fisheries management
EEZ	Exclusive economic zone
EU	European Union
FAO	Food and Agriculture Organization
ITQ	Individual transferable quota
IUU	Illegal, unreported and unregulated
MPA	Marine protected area
PNG	Papua New Guinea
SLA	Sustainable livelihoods approach
SPC	Secretariat of the Pacific Community
TAC	Total allowable catch
UNCLOS	United Nations Convention on the Law of the Sea
US	United States

LIST OF FIGURES

LIST OF TABLES

CHAPTER 1

Fishing Livelihoods and Fisheries Governance

Abstract This book centres on an understanding of fishing livelihoods within processes of historical change, and the social and political relationships within which they are embedded. Drawing on our research experience from the Asia-Pacific region, we examine where fishing livelihoods have come from, and where they are going. This introductory chapter introduces fishing livelihoods and the governance challenge that they face, before examining social science research in greater depth. We then develop the idea of a relational approach to fishing livelihoods, describing how they are shaped by wider political and economic trajectories, by local social relationships and by institutional structures.

Keywords Fishing livelihood • Political ecology • Wellbeing • Fisheries governance

In recent years the oceans have been subject to a profusion of regulatory, academic and private sector attention, as calls for a 'blue economy' are envisioned and executed across the world (Jouffray et al., 2020; Voyer et al., 2018). Characterised as the 'last frontier', oceans are presented as a dual opportunity for new forms of economic exploitation and renewed efforts to sustain ecological systems. Fisheries, and the livelihoods that they support, sit in an uneasy relationship to these transformations. While

1

fishing[1] has for generations provided food and livelihoods for millions of people throughout the world, increasingly it is challenged by newer coastal and ocean-based economic activities such as tourism and energy extraction, and by progressively tightening forms of governance that seek to reduce its environmental effects. The consequences of such developments are felt in different ways across the diverse social groups involved in fishing.

This book centres on an understanding of fishing livelihoods within processes of historical change, and the social and political relationships within which they are embedded. Drawing on our research experience from the Asia-Pacific region, we examine where fishing livelihoods have come from, and where they are going. Developing a 'relational' view of fishing livelihoods, we examine how they are shaped by wider political and economic trajectories, by local social relationships and by institutional structures. We discuss how such an understanding of fishing livelihoods can contribute to more ecologically sustainable and socially equitable governance strategies.

FISHING LIVELIHOODS

Across the world, fisheries provide direct employment for around 38.98[2] million people (Food and Agriculture Organization [FAO], 2020). In many coastal regions of the world, and particularly in many low-income contexts, fishing livelihoods remain the primary economic activity. Globally, they make significant contributions towards food and nutrition security, and are particularly important as a source of micronutrients, including vitamin A, omega-3, zinc, iron, calcium and selenium (Hicks et al., 2019).

Fishing livelihoods are characterised by their diversity, flexibility and dynamism, responding to changing environmental, climatic and economic conditions. It is common to differentiate between small-scale and large-scale fisheries, yet there are no universally accepted criteria that distinguish between these sectors. Large-scale fisheries tend to involve larger-sized vessels that use advanced or capital-intensive technologies (e.g., trawls, purse seines), wage labour and larger firms. The livelihoods in large-scale fisheries are as employed crew, or crew who are paid a portion of the value of the catch. In contrast, small-scale fisheries tend to be more labour

---

[1] By 'fish' we mean all seafood (e.g., including crustaceans, shellfish, etc.) in addition to fish. We do not include discussion of inland fishing or fishing livelihoods in this book.

[2] This includes part-time, seasonal and permanent work.

intensive and involve the use of smaller vessels and less capital-intensive gears (e.g., handlines, nets that can be pulled in by hand) that operate closer to shore, and are operated by individuals, households or small groups from within coastal villages. While small-scale fishers tend to sell their fish in local or domestic markets, or consume it directly, many are also involved in export operations. For example, Indonesian fishers who work alone from wooden vessels only a couple of metres long using hand-lines catch yellowfin tuna sold as steaks in North America. In practice, the distinctions between small-scale and large-scale fisheries blur considerably (Johnson et al., 2005). Small-scale fisheries are estimated to account for more than 90 per cent of fishers and fish workers (i.e., in trading and pro-cessing) (Kelleher et al., 2012), the vast majority of which are located in developing countries. While fishing is commonly thought of in relation to marine spaces, inland fisheries (the 'forgotten fisheries'; Cooke et al., 2016) account for about 12.5 per cent of total capture fisheries produc-tion (Funge-Smith & Bennett, 2019).

Beyond work in fishing alone, many livelihoods (over 100 million in total; Kelleher et al., 2012) are based around, partly or in whole, the diverse activities along the value chain—seafood processing, marketing, trading, boat and gear construction, servicing vessels and so on. Much work goes into preparing and maintaining the inputs for fishing, especially in boat and gear construction and maintenance. Post-harvest, seafood is processed in various ways. In large-scale fisheries, such as for tuna, pro-cessing plants and factories employ workers to prepare the fish according to different market demands, such as loins, cans or fillets. For example, in North-Eastern China, vast numbers of whitefish from Russia and other northern economies are imported, filleted and packaged, and then re-exported (Clarke, 2009). In small-scale fisheries, where access to refrigera-tion is less common, fish are commonly dried, smoked and/or salted.

Fish is one of the most highly traded commodities (Gephart & Pace, 2015), and this trade occurs at multiple scales. From local markets in vil-lages, to provincial town centres, to major national markets and overseas, fish are typically transported over land, sea and air through complex net-works of traders, agents and firms. Internationally, fish are exported to major markets including China, Japan, the European Union (EU) and North America, frequently through complex trade routes and intermedi-ary countries that make it difficult to track (Stoll et al., 2018). Wholesalers, retailers and restaurants then provide livelihoods for further nodes along fisheries value chains.

The current state of fisheries and their associated value chains and livelihoods is not a static picture, but is a reflection of the wider historical conditions that have led to this point. Fishing of various types has been prevalent for millennia, with political and economic developments shaping its character differently in different times and places. For example, the expansion of Chinese trade networks from the late seventeenth century led to the rapid uptake of fishing for products for the Chinese market, such as dried sea cucumbers. By the mid-nineteenth century, polities such as the Sulu Sultanate were centred on this economy (Warren, 1981). As domestic markets for fish products grew with increased populations and consumer demand, this stimulated the rise of specialised fishing communities along coastal South-East Asia (Spoehr, 1984). The foundations for industrial fisheries in South-East Asia were laid in the late 1800s and early twentieth century under European rule, and then industrial fisheries were established under Japanese imperial expansion southward in the decades preceding World War II (Butcher, 2004; Chen, 2008).

From the second half of the twentieth century, the intensification of globalisation brought dramatic changes to fishing livelihoods (Butcher, 2004). Increasing demand for fish in markets such as the EU and the United States (US) stimulated production, while new fishing technologies emerged to increase the efficiency of catch. Fisheries expanded geographically into new frontiers and intensified in locations where they had already been present, including deeper down the water column. In many locations, fishing livelihoods became transformed into a market-oriented activity based on trade at local, national and global scales. Globalisation, or the process of 'time–space compression', as Harvey (1989) terms it, has increased the scale, pace and diversity of fishing activities around the world (Eriksson et al., 2015; Gephart & Pace, 2015).

Fishing livelihoods are not simply an economic process of harvesting, processing and trading to generate income. Such economic activities are embedded within (Granovetter, 1985) and intersect with diverse social relationships. The particular manner by which fishing livelihoods are operated reflects wider social structures, which vary tremendously across geography and over time. For example, in many societies, fishing livelihoods are strongly gendered—men are frequently associated with fishing from boats further from shore, and women with near-shore fishing, gleaning, processing and marketing. Group identities such as ethnicity, caste, migration or religion or status can influence who is involved in fishing. For example, in South-East Asia, Sama-Bajau people are strongly associated with fishing practices. This association is often pejorative, and they are typecast as ignorant, poor and environmentally destructive (Lowe, 2000). In the Philippines, migrants from the Visayan group of islands are closely associated with fishing in some parts of the country where they lacked access to land for farming (Eder, 2003), and seafood exporters tend to have ethnic Chinese links. Many of the 'Japanese' fishers who worked throughout Asia and northern Pacific Island countries in the first half of the twentieth century were Okinawan, who through the forcible incorporation of Okinawa into the Japanese Empire in the 1870s were left destitute and had to travel to find livelihoods (Tomiyama, 2002).

The Governance Challenge

The environmental consequences of the progressive expansion and intensification of fisheries around the world have been significant. While there is much variability (Hilborn et al., 2020), in many cases, fisheries stocks have been overexploited. In 2017, the fraction of fish stocks considered by the FAO (2020) to be unsustainably fished was 34.2 per cent. Between 1950 and 2015, the catch-per-unit effort decreased by over 80 per cent in most countries (Rousseau et al., 2019). In many cases, the very viability of fishing livelihoods is under threat, following the trajectory of North American cod fisheries (Binkley, 2002).

Beyond the fish themselves, fish habitat and broader ecosystems have been substantially degraded. Destructive fishing gears such as dynamite, pollution, plastic debris, coastal infrastructure, shipping and agricultural run-off have negatively affected many marine ecosystems. Increasingly, the effects of climate change are being felt. The decline of coral reef systems, such as the Great Barrier Reef, is being driven primarily by coral bleaching caused by climate change (Hughes et al., 2017). Under a scenario of continued high emissions, the maximum catch potential of tropical fish stocks in some tropical exclusive economic zones (EEZs) is projected to decline by up to 40 per cent (Lam et al., 2020). Climate change is also projected to dramatically alter marine ecosystems through additional stresses such as ocean acidification, deoxygenation, changing patterns of nutrient supply and storms (Henson et al., 2017).

In this context of environmental decline, fishing livelihoods have been increasingly subject to attempts to govern their nature and extent. A central concern of much governance has focused on the need to sustainably manage fisheries as natural resources. At the local level, systems of customary marine tenure have regulated access to fishing grounds in many places (Foale et al., 2011). At a higher scale, state-based governance regimes have increasingly aimed to manage and regulate capture fisheries and the marine spaces where they are found (Campling & Havice, 2018). Through the twentieth century states exerted national claims over territorial waters (usually 12 nautical miles), and progressively expanded their claims over greater distances from the land. The United Nations Convention on the Law of the Sea (UNCLOS) emerged to codify sovereignty over EEZs in the early 1980s, and formally came into force from 1994, demarcating sovereign rights over waters 200 nm out from the coast.

In many richer countries, fisheries have been progressively managed through the use of economic instruments, founded on concepts such as maximum sustainable yield and maximum economic yield. Tools such as total allowable catches (TACs), licences and quotas are employed to regulate access to resources according to biologically determined parameters. Such instruments intersect with other regulations, including seasonal or other temporal closures, or gear restrictions. In recent decades, the concept of ecosystem-based fisheries management (EBFM) has become widely accepted, where fisheries are managed not as a single, isolated stock but with reference to the broader ecosystem of which they are part (Pikitch et al., 2004). Various forms of marine protected areas (MPAs) that spatially regulate access marine zones have become widespread as part of this.

Since the 2000s, the rise of market-based 'private governance' through various forms of certification and eco-labelling has become more prevalent (Bush & Oosterveer, 2019). The largest eco-label, the Marine Stewardship Council (MSC), now covers approximately 15 per cent of global fish catch (Le Manach et al., 2020). As consumers have become more aware of overfishing as a problem, in large part due to campaigns by environmental NGOs, some brands and retailers have sought to enhance their reputation by aligning themselves with the sustainable seafood movement. Others have used market-based initiatives to protect their reputation by avoiding association with destructive fishing practices.

More recently, a dominant governance paradigm appears to be coalescing around the idea of a 'blue economy'. While there is much variation in how this term is used (Silver et al., 2015), the core proposal is to manage marine resources in a way that integrates ecological sustainability and economic profitability. As economic and political actors increasingly seek to access marine resources for ecologically sustainable uses that generate economic value, fisheries are becoming pressured both by governance regimes and by newer sectors. In some cases, fishing livelihoods are being pushed out, as coastal land and waters are appropriated for tourism, energy and conservation (Barbesgaard, 2018; Bavinck et al., 2017).

All too commonly, much fisheries governance struggles to be effective in protecting both the fish stocks and the livelihoods that rely on them. Fishing itself is inherently difficult to govern because the act of fishing takes place out at sea. The challenge is particularly acute in low-income contexts, where the state has few resources available for management, implementation and enforcement, and in many instances there are few viable alternative livelihoods (Barclay et al., 2019). Small-scale fisheries typically tend to be widely dispersed and fragmented across coastlines, and in practice tend to be informally governed at more local scales (Steenbergen et al., 2019). They are frequently missed in formal state statistics, and as a result are less visible in policy—a state of affairs several organisations and programmes are aiming to change (e.g., Too Big to Ignore, 2013).

Yet, there are also more fundamental, underlying reasons behind the challenges of fisheries governance that go well beyond the technical challenges of resources, implementation and enforcement in remote locations. Much fisheries governance proceeds from a standpoint informed by a narrow set of perspectives, where livelihoods are understood in terms of how much dollar value fisheries generate, and/or the effects of fisheries on fish stocks and marine habitats. Financial revenue and the volume of landed catch become the two crucial metrics by which fisheries are assessed, and the role of fisheries management is usually framed in legislation as promoting viable industries and looking after fish stocks.

The consequences of such perspectives are twofold. First, they narrow our understanding of what a fishing livelihood is: they are 'reductionist' in that they reduce or limit the scope of understanding a fishing livelihood to the acts of fishing and selling, and indeed usually only to the formal, easily visible elements of seafood trade. As such, they tend to gloss over or miss key aspects of fishing livelihoods. The political and economic contexts of fishing livelihoods (including their integration with other forms of economic activity), the historical processes leading up to their contemporary configurations, and the diversity of social practices and identities associated with them are all central components of a fishing livelihood. As Johnson et al. (2005: 84) note, 'when looking at capture fisheries as a livelihood it becomes apparent that a strict division between the taking and landing of fish and other aspects of life is hard to maintain'.

Second, and relatedly, viewing fishing livelihoods through a lens that emphasises gross economic value and environmental effects leads to the formation of particular forms of governance that do not address some of these important aspects of fishing livelihoods. For example, small-scale fisheries tend to be more associated with marginalised groups, such as women or poorer people. The livelihood functions of small-scale fisheries can, therefore, be characterised more effectively through the generation of 'welfare', in contrast or in addition to 'wealth' (Béné et al., 2010). In these contexts, policies that aim to promote the generation of wealth alone can cause significant negative social effects (Cohen et al., 2019). Beyond the ethical dilemmas of these negative social effects, there are also environmentally pragmatic consequences of pursuing governance visions that ignore fishing livelihood contexts. When policies do not attain broad popular support, they fail to attain legitimacy (Coulthard et al., 2011; Jentoft, 2000). Governance that is illegitimate can have poor compliance. By ignoring the wider aspects of fishing livelihoods, governance is less likely to be legitimate, and subsequently less likely to attain its objectives.

FISHERIES AS A SOCIAL PROCESS

In contrast to perspectives that emphasise gross economic value and environmental effects of fisheries, there is a long tradition of social science that views fisheries as a fundamentally social process. Historians, social anthropologists and others have shown the intricate links between fishing and societies (Binkley, 2002; Clark, 2017; Firth, 1966; Probyn, 2016), documenting cultural traditions related to fishing (Allison et al., 2020), customary and contemporary forms of marine tenure that regulate access to marine spaces (Acheson, 1988; Hviding, 1996), the social relations between different groups of people involved in fishing (Pálsson, 1994), and the non-economic factors that drive people to pursue fishing as a livelihood (Pollnac & Poggie, 2008). In recent decades, much fisheries social science has taken an explicitly applied approach, seeking to apply insights about human behaviour to the challenge of improving fisheries governance (Berkes et al., 2001; Kooiman et al., 2005; Kraan & Linke, 2020; McGoodwin, 1995). Organisations such as the Centre for Maritime Studies, FAO, WorldFish and the Too Big to Ignore research network have developed significant bodies of literature around small-scale fisheries (Jentoft, 2019), the theory and practice of interactive governance (Kooiman et al., 2005) and human rights-based approaches to fisheries

(Allison et al., 2012). Many donor-funded fisheries projects in developing countries routinely include social science as part of their activities (e.g., Christie et al., 2005), and academic fields emerging from the environmental sciences (e.g., literature on social-ecological systems and resilience) now engage with questions traditionally addressed by social scientists, such as those relating to poverty and participation (e.g., Blythe et al., 2017). The increasing emphasis on interdisciplinary and multidisciplinary science has meant that the field of 'marine social science' (#marsocsci) now incorporates a diverse set of perspectives.

Within this field of scholarship, we focus on three specific threads of literature that are particularly relevant for our discussion on fishing livelihoods: the sustainable livelihoods approach (SLA), the social wellbeing approach and political economy perspectives. The SLA conceived of a livelihood comprised of 'the capabilities, assets (including both material and social resources) and activities required for a means of living' (Chambers & Conway, 1992: 6). Subsequently highly influential, in part because of its adoption by the United Kingdom's Department for International Development, livelihood assets were conceptualised as a 'pentagon' of capitals (natural, social, human, physical and financial). Among the key emphases of the SLA was a focus on diversification as a positive strategy to spread risk, and the SLA approach was subsequently used in the fisheries sector in diverse academic and policy contexts (Allison & Ellis, 2001; Allison & Horemans, 2006). While the SLA has subsequently been subject to critique for its relative neglect of politics and power (De Haan & Zoomers, 2005; Scoones, 2009), its emphasis on the material aspects of livelihoods (Carr, 2013) and on local-scale processes and structure (Carr, 2015), it remains a common approach in many fisheries governance interventions in developing countries (e.g., Apine et al., 2019).

Building on a diverse set of traditions in development studies and quality-of-life studies—including the SLA—the 'three-dimensional' or 3D social wellbeing approach emerged in fisheries as a way to understand fishing livelihoods more broadly than gross economic totals (Coulthard et al., 2011; Weeratunge et al., 2014). While 'objective' values (e.g., economic contributions) are still examined in this approach, attention is also paid to subjective values (e.g., job satisfaction) and relational values (e.g., relationships between different groups of people involved in the fishery). The goal is to assess a fishery not just in terms of economic value or environmental impact, but also in terms of a wider suite of values (Johnson et al., 2018). With this framing of fisheries, the governance question becomes one of how to design interventions that adequately capture the wider total of contributions that fisheries make (Song, 2018). Therefore, both the SLA and wellbeing approaches to understanding livelihoods take us a long way from understanding a fishing livelihood as the act of fish harvesting alone.

In the fisheries sector, much work in the broad tradition of political economy has emerged that challenges conventional explanations of resource decline as the 'tragedy of the commons', referring instead to wider systemic factors such as patterns of capital accumulation (Campling et al., 2012; Longo et al., 2015; Mansfield, 2004). However, as Belton (2016) notes, there remains considerable scope to bring together studies of social wellbeing and political economy. While studies of agrarian change that investigate the drivers and outcomes of livelihood change have long been prominent in journals such as the *Journal of Peasant Studies* and the *Journal of Agrarian Change*, less political economy attention has been paid to coastal spaces (Campling & Colás, 2018; Fabinyi et al., 2019).

OUR APPROACH

Focusing on the intersection between these approaches—livelihoods, wellbeing and political economy—our approach can broadly be seen as fitting in under the rubric of 'political ecology' (Perreault et al., 2015). This is a field notable for its diversity of concepts and approaches, but there are several key aspects of political ecology that inform our approach.

First is an emphasis on multiple scales. While recognising that scale itself is a social construct (Neumann, 2009), a core tenet of early political ecology from the 1980s has been to emphasise that the factors driving human behaviour in relation to the environment are often located at regional or global scales (Blaikie & Brookfield, 1987). This emphasis intersected with the long-overdue recognition in social anthropology that the social relations structuring everyday life frequently had as much or more to do with dynamic processes of global economic transformation over time, instead of what were typically depicted as static local cultures (Gupta & Ferguson, 1997; Wolf, 1983). Carr (2015: 336) notes that 'a political ecological approach to livelihoods analysis explains local livelihood decisions and their sustainability through locally specific materializations of translocal economic, political, and environmental processes and structures'.

A second notable feature of political ecology from at least the 1990s has been to analyse how material and symbolic orders interact (Hornborg et al., 2013; Peet & Watts, 1996). Social identities, ideas and cultural values—such as the role of gender (Rocheleau et al., 1996), attitudes towards the environment (Agrawal, 2005), or the roles of dominant environmental narratives (Fairhead & Leach, 1996; Forsyth, 2014)—all strongly influence human–environment relations. Relatedly, a third feature of much political ecology has been attention to the role of political actors, including the state, and political struggles over the environment (Robbins, 2012). From this perspective, conflicts over access to and exclusion from environmental resources are not unusual or aberrant processes, but the norm when studying 'politicised environments' (Bryant & Bailey, 1997; Le Billon & Duffy, 2018).

This book adopts what we term as a *relational approach* to fishing livelihoods. While the concept of relationality is used in diverse ways, here we draw broadly on a philosophical tradition that emphasises the roles of process, experience and relations as fundamental categories (Dewey, 1929; Ingold, 2015, 2018b; Whitehead, 1929). As Ingold (2018a: 100, 101–102) describes it, this is a view that sees 'relations not just as derivative of society, but as the very fabric of social life. ... In life, relations are not given in advance but have continually to be performed'. We use the term relational to emphasise that a livelihood is best understood as a set of activities operating in relationships with other processes and people over time, and that livelihoods are shaped by people's relational positions in society.

While closely related to the social wellbeing concept of relationality (see Johnson, 2018), we also draw explicitly on a political economy tradition of work on poverty. Specifically, we draw on Mosse's (2010) conceptualisation of poverty as a consequence of two sets of social relationships—first, historically developed economic and political relations, and second, social categorisation and identity (see also Harriss, 2009). While a 'fishing livelihood' is by no means always a life of poverty (Bavinck, 2014), and the concept of livelihood is quite different to the concept of poverty, we suggest that they are similar in that both can be effectively understood as centred on a set of social relationships that change over time, instead of as a discrete attribute of an individual or household.

In addition to the wider processes of political-economic change and the microsocial relations highlighted by Mosse, we suggest that a third key relationship a fishing livelihood has is with the specific institutional arrangements that govern access to and exclusion from fisheries resources (Hall et al., 2011; Li, 2007). While these three sets of relationships overlap with each other, the aim is to combine what are broadly Marxian ideas about 'adverse incorporation' into the global economy (e.g., McCarthy, 2010), with broadly Weberian ideas about social exclusion (e.g., Hall et al., 2011) and critical accounts of governance (Li, 2007) in complementary ways (Mosse, 2010). Understanding a livelihood in terms of the social relationships and structures that sustain and reproduce it embeds the concept in processes of change. We view livelihoods as constituted through their relations with the wider political economy, the microsocial climate and the institutional context.

The Asia-Pacific is an important site to study fishing livelihoods for several reasons. Asia alone provides 30.77 million out of the 38.98 million employed in fisheries worldwide (FAO, 2020). While the Pacific is far less densely populated, it generates some of the most globally significant fisheries in the form of large-scale tuna fisheries. The area as a whole is host to the 'Coral Triangle', a region defined by the highest marine biodiversity in the world. While we draw on secondary literature where relevant, the book draws directly on our own research experience across several countries. Fabinyi has conducted long-term research on fishing and coastal livelihoods in the Philippines since 2005, especially in Palawan and Mindoro.[3] He also has research experience on fishing livelihoods in Indonesia, Malaysia, Papua New Guinea (PNG) and Solomon Islands, and on seafood markets and consumption in China. Barclay started researching livelihoods in the tuna industry in Japan and the Solomon Islands in 1997, other Pacific Island countries, including PNG since 2005, and Indonesia since 2016. She has also investigated fisheries and aquaculture livelihoods in Australia (in South Australia, New South Wales, Victoria and Queensland). Both of us rely primarily on qualitative research, at different times using semi-structured interviews, participant observation and historical analysis. Since we draw on selected case studies from the countries where we have worked, the book is not a comprehensive analysis of the entire Asia-Pacific, but focuses largely on livelihoods in Island South-East Asia and the Pacific (see Fig. 1.1). Another major omission is of inland

---

[3] Mindoro is composed of two provinces, Occidental and Oriental Mindoro, while Palawan is currently proposed to be split into three provinces: Palawan del Norte, Palawan Oriental and Palawan del Sur. In this book, Mindoro and Palawan are collectively referred to as the Western Philippines.

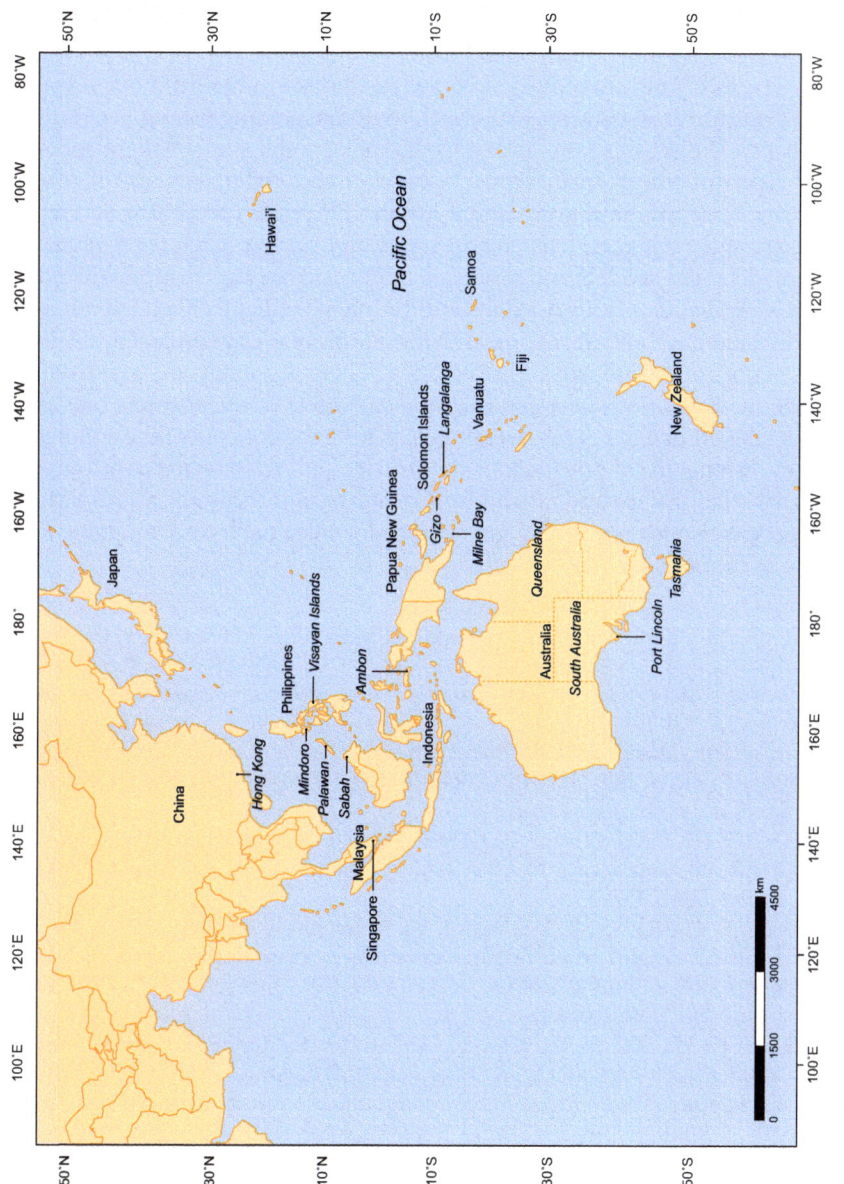

Fig. 1.1 Map of the Asia-Pacific

fishing livelihoods—an important topic deserving of greater levels of research and policy investment, particularly in the Asia-Pacific (Cooke et al., 2016; Funge-Smith & Bennett, 2019).

The following chapters draw on selected case studies from our research to demonstrate a relational approach to understanding fishing livelihoods. Chapter 2 discusses how fishing livelihoods are shaped by wider processes of capitalist transformation, using cases of the Philippines and PNG. In Chap. 3 we examine how fishing livelihoods relate to social processes of access and exclusion, particularly status and gender. Chapter 4 discusses how different models and practices of governance can shape livelihoods, drawing on cases from Australia and Indonesia. Chapter 5 concludes with a discussion of how the approach taken in this book can be practically used to contribute to improved governance. There are many areas of overlap, and the distinctions between the subject matters of the chapters blur considerably in practice. Our overall goal is to highlight concrete examples of how fishing livelihoods relate to broader political-economic processes, social relationships and institutional contexts, and the implications of such a perspective for improving governance for sustainable and equitable fishing livelihoods.

References

Acheson, J. M. (1988). *The lobster gangs of Maine*. University Press of New England.

Agrawal, A. (2005). *Environmentality: Technologies of government and the making of subjects*. Duke University Press.

Allison, E. H., & Ellis, F. (2001). The livelihoods approach and management of small-scale fisheries. *Marine Policy, 25*(5), 377–388.

Allison, E. H., & Horemans, B. (2006). Putting the principles of the sustainable livelihoods approach into fisheries development policy and practice. *Marine Policy, 30*(6), 757–766.

Allison, E. H., Ratner, B. D., Åsgård, B., Willmann, R., Pomeroy, R., & Kurien, J. (2012). Rights-based fisheries governance: From fishing rights to human rights. *Fish and Fisheries, 13*(1), 14–29. https://doi.org/10.1111/j.1467-2979.2011.00405.x

Allison, E. H., Kurien, J., Ota, Y., Adhuri, D. S., Bavinck, J. M., Cisneros-Montemayor, A., Olukoju, A., et al. (2020). The human relationship with our ocean planet. Washington, DC: World Resources Institute. Retrieved December 10, 2020, from https://www.oceanpanel.org/blue-papers/Human RelationshipwithOurOceanPlanet

Apine, E., Turner, L. M., Rodwell, L. D., & Bhatta, R. (2019). The application of the sustainable livelihood approach to small scale-fisheries: The case of mud crab *Scylla serrata* in south West India. *Ocean & Coastal Management, 170*, 17–28.

Barbesgaard, M. (2018). Blue growth: Savior or ocean grabbing? *The Journal of Peasant Studies, 45*(1), 130–149. https://doi.org/10.1080/0306615 0.2017.1377186

Barclay, K., Fabinyi, M., Kinch, J., & Foale, S. (2019). Governability of high-value fisheries in low-income contexts: A case study of the sea cucumber fishery in Papua New Guinea. *Human Ecology, 47*(3), 381–396. https://doi. org/10.1007/s10745-019-00078-8

Bavinck, M. (2014). Investigating poverty through the lens of riches—Immigration and segregation in Indian capture fisheries. *Development Policy Review, 32*(1), 33–52. https://doi.org/10.1111/dpr.12042

Bavinck, M., Berkes, F., Charles, A. T., Dias, A. C. E., Doubleday, N., Nayak, P. K., & Sowman, M. (2017). The impact of coastal grabbing on community conservation—A global reconnaissance. *Maritime Studies, 16*(1), 8. https:// doi.org/10.1186/s40152-017-0062-8

Belton, B. (2016). Shrimp, prawn and the political economy of social wellbeing in rural Bangladesh. *Journal of Rural Studies, 45*, 230–242.

Béné, C., Hersoug, B., & Allison, E. H. (2010). Not by rent alone: Analysing the pro-poor functions of small-scale fisheries in developing countries. *Development Policy Review, 28*(3), 325–358. https://doi.org/10.1111/j.1467-7679. 2010.00486.x

Berkes, F., Mahon, R., McConney, P., Pollnac, R., & Pomeroy, R. (2001). *Managing small-scale fisheries: Alternative directions and methods*. IDRC.

Binkley, M. E. (2002). *Set adrift: Fishing families*. University of Toronto Press.

Blaikie, P., & Brookfield, H. (Eds.). (1987). *Land degradation and society* (1st ed.). Routledge.

Blythe, J., Cohen, P. J., Eriksson, H. B., Cinner, J. E., Boso, D., Schwarz, A.-M., & Andrew, N. L. (2017). Strengthening post-hoc analysis of community-based fisheries management through the social-ecological systems framework. *Marine Policy, 82*, 50–58.

Bryant, R. L., & Bailey, S. (1997). *Third world political ecology*. Routledge.

Bush, S. R., & Oosterveer, P. (2019). *Governing sustainable seafood*. Routledge.

Butcher, J. G. (2004). *The closing of the frontier: A history of the marine fisheries of Southeast Asia, c. 1850–2000*. Institute of Southeast Asian Studies.

Campling, L., & Colás, A. (2018). Capitalism and the sea: Sovereignty, territory and appropriation in the global ocean. *Environment and Planning D: Society and Space, 36*(4), 776–794. https://doi.org/10.1177/0263775817737319

Campling, L., & Havice, E. (2018). The global environmental politics and political economy of seafood systems. *Global Environmental Politics, 18*(2), 72–92. https://doi.org/10.1162/glep_a_00453

Campling, L., Havice, E., & McCall Howard, P. (2012). The political economy and ecology of capture fisheries: Market dynamics, resource access and relations of exploitation and resistance. *Journal of Agrarian Change, 12*(2–3), 177–203. https://doi.org/10.1111/j.1471-0366.2011.00356.x

Carr, E. R. (2013). Livelihoods as intimate government: Reframing the logic of livelihoods for development. *Third World Quarterly, 34*(1), 77–108. https://doi.org/10.1080/01436597.2012.755012

Carr, E. R. (2015). Political ecology and livelihoods. In T. Perreault, G. Bridge, & J. McCarthy (Eds.), *The Routledge handbook of political ecology* (pp. 332–342). Routledge.

Chambers, R., & Conway, G. (1992). *Sustainable rural livelihoods: Practical concepts for the 21st century.* IDS Discussion Paper no. 296. Institute of Development Studies.

Chen, H. (2008). Japan and the birth of Takao's fisheries in Nanyo, 1895–1945. *International Journal of Maritime History, 20*(1), 133–152. https://doi.org/10.1177/084387140802000107

Christie, P., Lowry, K., White, A. T., Oracion, E. G., Sievanen, L., Pomeroy, R., Pollnac, R., et al. (2005). Key findings from a multidisciplinary examination of integrated coastal management process sustainability. *Ocean & Coastal Management, 48*(3–6), 468–483. https://doi.org/10.1016/j.ocecoaman.2005.04.006

Clark, A. (2017). *The catch: The story of fishing in Australia* (1st ed.). National Library of Australia.

Clarke, S. (2009). *Understanding China's fish trade and traceability.* TRAFFIC East Asia.

Cohen, P. J., Allison, E. H., Andrew, N. L., Cinner, J., Evans, L. S., Fabinyi, M., Jentoft, S., et al. (2019). Securing a just space for small-scale fisheries in the blue economy. *Frontiers in Marine Science, 6,* 171. https://doi.org/10.3389/fmars.2019.00171

Cooke, S. J., Allison, E. H., Beard, T. D., Jr., Arlinghaus, R., Arthington, A. H., Bartley, D. M., Cowx, I. G., et al. (2016). On the sustainability of inland fisheries: Finding a future for the forgotten. *Ambio, 45*(7), 753–764. https://doi.org/10.1007/s13280-016-0787-4

Coulthard, S., Johnson, D., & McGregor, J. A. (2011). Poverty, sustainability and human wellbeing: A social wellbeing approach to the global fisheries crisis. *Global Environmental Change, 21*(2), 453–463.

De Haan, L., & Zoomers, A. (2005). Exploring the frontier of livelihoods research. *Development and Change, 36*(1), 27–47. https://doi.org/10.1111/j.0012-155X.2005.00401.x

Dewey, J. (1929). *Experience and nature.* Open Court Publishing.

Eder, J. F. (2003). Of fishers and farmers: Ethnicity and resource use in coastal Palawan. *Philippine Quarterly of Culture and Society, 31*(3), 207–225.

Eriksson, H., Österblom, H., Crona, B., Troell, M., Andrew, N., Wilen, J., & Folke, C. (2015). Contagious exploitation of marine resources. *Frontiers in Ecology and the Environment, 13*(8), 435–440. https://doi.org/ 10.1890/140312

Fabinyi, M., Dressler, W., & Pido, M. (2019). Access to fisheries in the maritime frontier of Palawan Province, Philippines. *Singapore Journal of Tropical Geography, 40*(1), 92–110.

Fairhead, J., & Leach, M. (1996). *Misreading the African landscape: Society and ecology in a forest-savanna mosaic* (Vol. 90). Cambridge University Press.

Firth, R. (1966). *Malay fishermen: Their peasant economy* (2nd ed.). Routledge & Kegan Paul.

Foale, S., Cohen, P., Januchowski-Hartley, S., Wenger, A., & Macintyre, M. (2011). Tenure and taboos: Origins and implications for fisheries in the Pacific. *Fish and Fisheries, 12*(4), 357–369. https://doi.org/10.1111/ j.1467-2979.2010.00395.x

Food and Agriculture Organization. (2020). *The state of world fisheries and aqua-culture 2020: Sustainability in action.* FAO. https://doi.org/10.4060/ca9229en

Forsyth, T. (2014). Public concerns about transboundary haze: A comparison of Indonesia, Singapore, and Malaysia. *Global Environmental Change, 25*, 76–86.

Funge-Smith, S., & Bennett, A. (2019). A fresh look at inland fisheries and their role in food security and livelihoods. *Fish and Fisheries, 20*(6), 1176–1195. https://doi.org/10.1111/faf.12403

Gephart, J. A., & Pace, M. L. (2015). Structure and evolution of the global sea-food trade network. *Environmental Research Letters, 10*(12), 125014.

Granovetter, M. (1985). Economic action and social structure: The problem of embeddedness. *American Journal of Sociology, 91*(3), 481–510. https://doi. org/10.1086/228311

Gupta, A., & Ferguson, J. (Eds.). (1997). *Culture, power, place: Explorations in critical anthropology.* Duke University Press.

Hall, D., Hirsch, P., & Li, T. M. (2011). *Powers of exclusion: Land dilemmas in Southeast Asia.* University of Hawai'i Press.

Harriss, J. (2009). Bringing politics back into poverty analysis. In T. Addison, D. Hulme, & R. Kanbur (Eds.), *Poverty dynamics: Interdisciplinary perspectives* (pp. 205–224). Oxford University Press.

Harvey, D. (1989). *The condition of postmodernity.* Blackwell.

Henson, S. A., Beaulieu, C., Ilyina, T., John, J. G., Long, M., Séférian, R., Sarmiento, J. L., et al. (2017). Rapid emergence of climate change in environmental drivers of marine ecosystems. *Nature Communications, 8*(1), 1–9. https://doi.org/10.1038/ncomms14682

Hicks, C. C., Cohen, P. J., Graham, N. A., Nash, K. L., Allison, E. H., D'Lima, C., MacNeil, M. A., et al. (2019). Harnessing global fisheries to tackle micro-nutrient deficiencies. *Nature, 574*(7776), 95–98. https://doi.org/10.1038/ s41586-019-1592-6

Hilborn, R., Amoroso, R. O., Anderson, C. M., Baum, J. K., Branch, T. A., Costello, C., Kurota, H., et al. (2020). Effective fisheries management instrumental in improving fish stock status. *Proceedings of the National Academy of Sciences, 117*(4), 2218–2224. https://doi.org/10.1073/pnas.1909726116

Hornborg, A., Clark, B., & Hermele, K. (2013). Introduction: Ecology and power. In A. Hornborg, B. Clark, & K. Hermele (Eds.), *Ecology and power: Struggles over land and material resources in the past, present and future* (Vol. 18). Routledge.

Hughes, T. P., Kerry, J. T., Álvarez-Noriega, M., Álvarez-Romero, J. G., Anderson, K. D., Baird, A. H., Bridge, T. C., et al. (2017). Global warming and recurrent mass bleaching of corals. *Nature, 543*(7645), 373–377. https://doi.org/10.1038/nature21707

Hviding, E. (1996). *Guardians of Marovo lagoon: Practice, place, and politics in maritime Melanesia* (Vol. 14). University of Hawaii Press.

Ingold, T. (2015). *The life of lines*. Routledge.

Ingold, T. (2018a). *Anthropology: Why it matters*. Polity Press.

Ingold, T. (2018b). Evolution in the minor key. In A. Fuentes & C. Deane-Drummond (Eds.), *Evolution of wisdom: Major and minor keys* (pp. 115–123). Center for Theology Science and Human Flourishing, University of Notre Dame.

Jentoft, S. (2000). Legitimacy and disappointment in fisheries management. *Marine Policy, 24*(2), 141–148.

Jentoft, S. (2019). Life above water: Essays on human experiences of small-scale fisheries. In *TBTI Global Book Series 1*. Global.

Johnson, D. S. (2018). The values of small-scale fisheries. In D. S. Johnson, T. G. Acott, N. Stacey, & J. Urquhart (Eds.), *Social wellbeing and the values of small-scale fisheries* (pp. 1–21). Springer.

Johnson, D. S., Bavinck, M., & Veitayaki, J. (2005). Fish capture. In J. Kooiman, M. Bavinck, S. Jentoft, & R. Pullin (Eds.), *Fish for life: Interactive governance for fisheries* (pp. 71–91). University of Amsterdam Press.

Johnson, D. S., Acott, T. G., Stacey, N., & Urquhart, J. (Eds.). (2018). *Social wellbeing and the values of small-scale fisheries*. Springer.

Jouffray, J.-B., Blasiak, R., Norström, A. V., Österblom, H., & Nyström, M. (2020). The blue acceleration: The trajectory of human expansion into the ocean. *One Earth, 2*(1), 43–54.

Kelleher, K., Westlund, L., Hoshino, E., Mills, D., Willmann, R., de Graaf, G., Brummett, R. (2012). *Hidden harvest: The global contribution of capture fisheries*. Report no. 66469-GLB. World Bank.

Kooiman, J., Jentoft, S., Bavinck, M., & Pullin, R. (Eds.). (2005). *Fish for life: Interactive governance for fisheries*. Amsterdam University Press.

Kraan, M., & Linke, S. (2020). Commentary 2 to the manifesto for the marine social sciences: Applied social science. *Maritime Studies, 19*(2), 129–130.

Lam, V. W., Allison, E. H., Bell, J. D., Blythe, J., Cheung, W. W., Frölicher, T. L., Sumaila, U. R., et al. (2020). Climate change, tropical fisheries and prospects for sustainable development. *Nature Reviews Earth & Environment*, *1*(9), 440–454.

Le Billon, P., & Duffy, R. V. (2018). Conflict ecologies: Connecting political ecology and peace and conflict studies. *Journal of Political Ecology, 25*(1), 239–260. https://doi.org/10.2458/v25i1.22704

Le Manach, F., Jacquet, J. L., Bailey, M., Jouanneau, C., & Nouvian, C. (2020). Small is beautiful, but large is certified: A comparison between fisheries the marine stewardship council (MSC) features in its promotional materials and MSC-certified fisheries. *PLoS One, 15*(5), e0231073. https://doi.org/10.1371/journal.pone.0231073

Li, T. M. (2007). *The will to improve: Governmentality, development, and the practice of politics*. Duke University Press.

Longo, S. B., Clausen, R., & Clark, B. (2015). *The tragedy of the commodity: Oceans, fisheries, and aquaculture*. Rutgers University Press.

Lowe, C. (2000). Global markets, local injustice in southeast Asian seas: The live fish trade and local fishers in the Togean Islands of Sulawesi. In C. Zerner (Ed.), *People, plants, and justice: The politics of nature conservation* (pp. 234–258). Columbia University Press.

Mansfield, B. (2004). Rules of privatization: Contradictions in neoliberal regulation of North Pacific fisheries. *Annals of the Association of American Geographers, 94*(3), 565–584. https://doi.org/10.1111/j.1467-8306.2004.00414.x

McCarthy, J. F. (2010). Processes of inclusion and adverse incorporation: Oil palm and agrarian change in Sumatra, Indonesia. *The Journal of Peasant Studies, 37*(4), 821–850. https://doi.org/10.1080/03066150.2010.512460

McGoodwin, J. R. (1995). *Crisis in the world's fisheries: People, problems, and policies*. Stanford University Press.

Mosse, D. (2010). A relational approach to durable poverty, inequality and power. *The Journal of Development Studies, 46*(7), 1156–1178. https://doi.org/10.1080/00220388.2010.487095

Neumann, R. P. (2009). Political ecology: Theorizing scale. *Progress in Human Geography, 33*(3), 398–406. https://doi.org/10.1177/0309132508096353

Pálsson, G. (1994). *Coastal economies, cultural accounts: Human ecology and Icelandic discourse*. Manchester University Press.

Peet, R., & Watts, M. (Eds.). (1996). *Liberation ecologies: environment, development, social movements*. Routledge.

Perreault, T., Bridge, G., & McCarthy, J. (Eds.). (2015). *The Routledge handbook of political ecology*. Routledge.

Pikitch, E. K., Santora, C., Babcock, E. A., Bakun, A., Bonfil, R., Conover, D. O., Houde, E. D., et al. (2004). Ecosystem-based fishery management. *Science, 305*(5682), 346–347. https://doi.org/10.1126/science.1098222

Pollnac, R. B., & Poggie, J. J. (2008). Happiness, well-being and psychocultural adaptation to the stresses associated with marine fishing. *Human Ecology Review, 15*(2), 194–200.

Probyn, E. (2016). *Eating the ocean.* Duke University Press.

Robbins, P. (2012). *Political ecology: A critical introduction* (Vol. 16, 2nd ed.). John Wiley & Sons.

Rocheleau, D., Thomas-Slayter, B., & Wangari, E. (Eds.). (1996). *Feminist political ecology: Global issues and local experience.* Routledge.

Rousseau, Y., Watson, R. A., Blanchard, J. L., & Fulton, E. A. (2019). Evolution of global marine fishing fleets and the response of fished resources. *Proceedings of the National Academy of Sciences, 116*(25), 12238–12243. https://doi.org/10.1073/pnas.1820344116

Scoones, I. (2009). Livelihoods perspectives and rural development. *The Journal of Peasant Studies, 36*(1), 171–196. https://doi.org/10.1080/03066150902820503

Silver, J. J., Gray, N. J., Campbell, L. M., Fairbanks, L. W., & Gruby, R. L. (2015). Blue economy and competing discourses in international oceans governance. *The Journal of Environment & Development, 24*(2), 135–160. https://doi.org/10.1177/1070496515580797

Song, A. M. (2018). How to capture small-scale fisheries' many contributions to society?—Introducing the 'value-contribution matrix' and applying it to the case of a swimming crab fishery in South Korea. In D. Johnson, T. Acott, N. Stacey, & J. Urquhart (Eds.), *Social wellbeing and the values of small-scale fisheries* (pp. 125–146). Springer.

Spoehr, A. (1984). Change in Philippine capture fisheries: An historical overview. *Philippine Quarterly of Culture and Society, 12*(1), 25–56.

Steenbergen, D. J., Fabinyi, M., Barclay, K., Song, A. M., Cohen, P. J., Eriksson, H., & Mills, D. J. (2019). Governance interactions in small-scale fisheries market chains: Examples from the Asia-Pacific. *Fish and Fisheries, 20*(4), 697–714. https://doi.org/10.1111/faf.12370

Stoll, J. S., Crona, B. I., Fabinyi, M., & Farr, E. R. (2018). Seafood trade routes for lobster obscure teleconnected vulnerabilities. *Frontiers in Marine Science, 5,* 239. https://doi.org/10.3389/fmars.2018.00239

Tomiyama, I. (2002). The 'Japanese' of Micronesia: Okinawans in the Nanyô Islands. In R. Y. Nakasone (Ed.), *Okinawan diaspora.* University of Hawai'i Press.

Too Big to Ignore. (2013). Too Big to Ignore. Retrieved February 5, 2021, from http://toobigtoignore.net/

Voyer, M., Quirk, G., McIlgorm, A., & Azmi, K. (2018). Shades of blue: What do competing interpretations of the blue economy mean for oceans governance? *Journal of Environmental Policy & Planning, 20*(5), 595–616. https://doi.org/10.1080/1523908X.2018.1473153

Warren, J. F. (1981). *The Sulu zone, 1768–1898: The dynamics of external trade, slavery, and ethnicity in the transformation of a southeast maritime state.* Singapore University Press.

Weeratunge, N., Béné, C., Siriwardane, R., Charles, A., Johnson, D., Allison, E. H., Badjeck, M. C., et al. (2014). Small-scale fisheries through the wellbeing lens. *Fish and Fisheries, 15*(2), 255–279. https://doi.org/10.1111/faf.12016

Whitehead, A. N. (1929). *Process and reality: An essay in cosmology.* Cambridge University Press.

Wolf, E. R. (1983). *Europe and the people without history.* University of California Press.

CHAPTER 2

Responding to Global Change

Abstract This chapter focuses on the wider processes of political-economic change that drive key characteristics of fishing livelihoods. Globalisation has dramatically expanded the scale and accelerated the pace of fisheries capture and trade, generating new opportunities and challenges for livelihoods and marine environments. Here we document some of the major characteristics of the history of fishing across the Asia-Pacific, before focusing on case studies of the Philippines and PNG. We highlight three related features of globalisation that have influenced fishing livelihoods and that continue to shape them today: migration, engagement with markets and new technologies, and interactions with other forms of economic activity, including those outside the fisheries sector.

Keywords Globalisation • Fish markets • Fishing technology • Papua New Guinea • Philippines

In this chapter we focus on the wider processes of political-economic change that drive key characteristics of fishing livelihoods. Along with other sectors of economic life, fisheries have been radically transformed through an interrelated set of processes commonly referred to by the shorthand term 'globalisation'. Globalisation has dramatically expanded

© The Author(s) 2022   23
M. Fabinyi, K. Barclay, *Asia-Pacific Fishing Livelihoods*,
https://doi.org/10.1007/978-3-030-79591-7_2

the scale and accelerated the pace of fisheries capture and trade, generating new opportunities and challenges for livelihoods and marine environments.

While analyses of the relationships between globalisation and agrarian livelihoods are common (Akram-Lodhi & Kay, 2010; Bernstein, 2010), less frequent are examinations of the influences of globalisation on fishing livelihoods. While not denying the capacity of human agency and choice that feeds into livelihood decisions (Ransan-Cooper, 2016), the livelihoods of fishers respond to regional and global forces that change over time. Whether it be crew on industrial fishing vessels (Minnegal et al., 2003; Pálsson & Durrenberger, 1990), or small-scale fishers accessing new markets, adopting new technologies or transitioning into different forms of production (e.g., aquaculture) (Belton & Thilsted, 2014; Béné et al., 2009; Platteau, 1984), the structural conditions of these economic activities are formed by processes of change that operate at much wider scales than the household or the local community.

Here we document some of the major characteristics of the history of fishing across the Asia-Pacific, before focusing on case studies of the Philippines and PNG. We highlight three related features of globalisation that have influenced fishing livelihoods and that continue to shape them today: migration, engagement with markets and new technologies, and interactions with other forms of economic activity, including those outside the fisheries sector.

FISHERIES AND GLOBALISATION

The term 'globalisation' has many interpretations. We adopt a perspective that views it as a process centred around capital accumulation. Harvey's concepts of the 'spatial fix' (1982) and of 'time–space compression' (1989) highlight the drive for capital accumulation, and the increasing power of communication and transport technologies. While this process arguably started with European capitalism and colonialism spreading out from the sixteenth century (Wallerstein, 2004), these processes have intensified since the end of World War II. From this perspective, globalisation is a systemic force driven by capitalism.

A key theme in historical accounts of the globalisation of fisheries has been that of the 'frontier'—fisheries activities expanding and intensifying in response to new market demands from population growth and increasing wealth (Butcher, 2004). Moore's (2015) commodity frontier framework distinguishes between two phases of frontiers, and has been usefully applied to tuna fisheries (Campling, 2012). The initial phase of 'commodity widening' is based on geographic expansion, and 'commodity deepening' involves intensive development: 'firms dependent upon the appropriation of natural resources seek to continuously expand into new commodity frontiers, whether in terms of geographical extent or industrial intensity' (Baglioni & Campling, 2017: 2443). Fishing livelihoods have been progressively drawn into these dynamic forces of capital accumulation—moving towards new opportunities; using new technologies to target more types of marine resources, preserve them more effectively and transport them more easily; and accessing trade networks operating at greater scales.

Butcher (2004) and Christensen (2014) distinguish three key phases of development in the commercial fisheries of South-East Asia and the 'Indo-Pacific', respectively. From the later part of the nineteenth century until the 1930s, the foundations for industrial fishing in South-East Asia were laid by European colonial powers bringing small island groups together into states, and establishing government control over coastal areas, reducing piracy, expanding transport networks and encouraging fishing investment (Butcher, 2004). In the early twentieth century Japanese fleets expanded industrial fishing into South-East Asia, and built smoking and canning factories (Butcher, 2004; H. Chen 2008a; T.Y. Chen 2008b; Fujinami, 1987). After World War II, fisheries activities in South-East Asia boomed in what was termed 'the great fish race' (Butcher, 2004) or the 'great acceleration' (Christensen, 2014). State-supported fisheries expanded rapidly, and in the 1950s pre-war Japanese interests also re-established industrial fishing enterprises in South-East Asia (Morgan & Staples, 2006). Subsequently, from the late 1970s, the frontier began to 'close', as fish catches began to stagnate.

As Campling and Havice (2018) note, national seafood production systems and state-based regulatory regimes have been a crucial element of this process. In the Asia-Pacific, one of the key factors at play has been the rise of different distant water fleets seeking catches around the globe. Fishing states have supported their fleets to fish in distant waters by various means, including subsidies for vessels, fuel and fisheries access fees. For their part, coastal states have shaped fishing patterns by excluding distant water fleets from their EEZs in favour of domestic fleets, as in the case of the Maldives. Kiribati and Vanuatu are examples of a different route, inviting distant water fleets to fish in their EEZs in exchange for fees. A third model is to invite distant water fishing companies to invest in domestic fishing and/or processing capacity, like PNG and Solomon Islands (Barclay & Cartwright, 2008). The Japanese and US industrial tuna fleets were the first in the Pacific, although, by the start of the 'great acceleration' both were waning somewhat, in part due to their own rising production costs compared to competitor fleets. Taiwanese and Korean fleets became important during the heyday expansion of the 1970s and 1980s. In the 2000s the Filipino tuna fleet became a regional player, and the Chinese fleet started its steady increase (Barclay, 2014).

As countries became exposed to fishing practices from other countries, knowledge diffused and technology developed. Harvesting practices adopted new technologies to increase their catch, such as nylon fibres for nets and engines for boats. For example, the 'muro-ami' fishing technique was introduced to the Philippines by Japan, and was adopted by many vessels in the Philippines until its eventual banning in 1986. Blast fishing became popular after World War II, when access to explosives became common. Processing and preservation technologies, such as canning and freezing, reduced the perishability of fish, and the expansion of transportation networks (e.g., airfreight) all contributed to greater capacity to store and distribute fish catch. The mixing of knowledge and technology is also closely linked to the increase of movement and people, as people moved to access the opportunities provided by fishing livelihoods in more productive places (Eder, 2008). Many industrial fisheries in the Asia-Pacific came to be crewed by foreign labour.

Specific consumer markets have also emerged as crucial drivers of the growth in fisheries. Much seafood exportation has been from developing to developed countries, especially Japan, the EU and the US (Swartz et al., 2010). Consumer preferences, such as for tuna in Japan, have shaped what sorts of fish are caught and how they are processed, and consumer markets increasingly shape the regulatory conditions under which fish are caught through trade measures. The growth of China as a wealthy consumer market has had significant effects on the nature of fishing livelihoods in the Asia-Pacific. While many of the products demanded by the Chinese market are not new, increasing wealth in China since the opening up of the economy in the 1980s has greatly increased demand. Markets for sea cucumbers, shark fin, live reef food fish and fish maw—products highly valued in Chinese cuisine for perceived health benefits or associations with high status—have all expanded greatly since this period (Purcell et al., 2013; Sadovy de Mitcheson et al., 2013; Sadovy de Mitcheson et al.,

2019; Scales et al., 2006). This has led analysts to conceptualise the nature of this form of seafood production and trade as 'contagious exploitation' (Eriksson et al., 2015), or as 'roving bandits' (Scales et al., 2006). More recently, COVID-19 is reshaping the nature of seafood markets and of trade more generally (Robins et al., 2020).

The following sections explore how these global-scale processes have come to concretely shape fishing livelihoods, focusing on three key aspects of globalisation: migration, new technologies and markets, and changing forms of interaction with non-fisheries activities.

The Western Philippines

A particular form of movement in the Philippines since the late nineteenth century has been individuals and families leaving locations characterised by social conflict, high population densities and poverty, towards 'frontier' locations characterised by lower population densities and new livelihood opportunities. At different points in time, Mindanao, Mindoro and Palawan served as such frontier settlement locations.

In Mindoro, Indigenous Mangyan groups were once the majority population, but this changed with settlement by migrants during the early twentieth century. They arrived from varied regions of the Philippines (e.g., Luzon, the Visayan Islands) in response to US colonial encouragement of agricultural production and exports, and investments in infrastructure (Helbling & Schult, 1997; Schult, 1991). Migrants settled heavily in the coastal and agriculturally productive lowlands of the island, while Mangyan groups became marginalised upland. Land conflicts soon ensued (Helbling & Schult, 1997). Migrants outnumbered the Mangyan by 1920 (1997) and came to dominate the lowlands and coasts, and consequently the fishing livelihoods.

In Palawan, Indigenous groups occupied different parts of the province prior to settlement by migrants—Tagbanua in the North, Batak and Palawan in the central part of the island, and Molbog in the South. From the late nineteenth century, migration from the nearby islands of Cuyo and Agutaya increased. Long characterised as the 'last frontier of the Philippines', migration to Palawan increased after the settlement of Mindoro and Mindanao, and particularly so after World War II. While migrants arrived from diverse locations, many came from the Visayan group of islands (Eder & Fernandez, 1996).

Compared to Mindoro the settlement of Palawan was less driven by specific projects, and Palawan instead served as a 'land of opportunity' for farming and fishing in particular. For example, fishers would often travel to Palawan on a seasonal or intermittent basis, sometimes forming social relationships with local groups already present (Ushijima & Zayas, 1994). After some time of these sojourns, the household might relocate, and then other kin and neighbours would follow (Seki, 2004). The time line of migration to Palawan—Cuyonon initially, followed by Visayan—meant that many of the best farmlands were obtained by Cuyonon households, and more recent Visayan migrants settled along the coast (Eder, 2003). In the south of Palawan, refugees from the civil conflict in the Sulu Archipelago settled from the 1970s.

While the Western Philippines was the source of a significant proportion of the country's entire landed catch from the 1970s (Butcher, 2004), the vast majority of the vessels were based in Manila and elsewhere in the Philippines, contributing little to the local economy. Although since 1991 the Local Government Code has prohibited the entry of large-scale fishing vessels[1] within 15 km of the shoreline, the waters around Mindoro and Palawan remain favoured fishing grounds for many externally based large-scale vessels. Over time, large-scale fisheries based in these locations also emerged. For example, in the 1980s Coron in Northern Palawan became host to a high number of lift net boats, targeting anchovies. Other commercial fishing boats adapted gears and techniques to fish for fusiliers with baited hook and line; mixed reef fish with weighted lines with lures; dynamite (illegal); spearfishing using air compressors for diving; variants of 'baby' purse seines and trawlers; and the notorious muro-ami fishery, notable for its high degree of ecological destructiveness and use of child labour. Due to the lack of readily available ice, processing and preserving fish through drying was very common. The fish landed from these vessels served major provincial markets throughout the country, and especially that of Manila, while some were exported.

Small-scale fishers in the Philippines have long been selling their catch commercially. As Firth (1966) and Spoehr (1984) noted, because fish alone does not provide an adequate source of food, full-time fishers (or specialised fishing communities) in particular needed to sell some of their catch to obtain other foods. Fish was bartered for rice, vegetables or other

---

[1] Large-scale fisheries in the Philippines are labelled as 'commercial fishing', defined as > 3 gross tonnes.

food, or sold at local markets for cash, or dried and sold to itinerant buyers and vessels that collected fish products. With the introduction of motorised boats, and the increased availability of ice and roads, the capacity of small-scale fishers to access new markets, such as the municipal towns, increased.

With the spread of migrants came the spread of knowledge about fishing techniques—migrants from the Visayan region are particularly renowned for their knowledge of different types of fishing gears and techniques. Small-scale fishers in contemporary times in the Western Philippines are notable for the extraordinary diversity of gears and techniques used to obtain marine resources (see Fig. 2.1). Simple hook and line, originally using vine or other natural fibres, was replaced with nylon, and different techniques for hooking fish include the use of bait (e.g., shrimp) and the use of lures (e.g., foil, plastic, feathers). Longlines, utilising many larger-sized hooks to a greater depth, are used to catch larger fish and sharks. Variants of net fishing including bottom-set, floating and drift gillnets are common, sometimes with the use of plungers to scare the fish into the net. Traps made of bamboo or other wood for fish and crustaceans are

Fig. 2.1 Fishers catching big-eye scad to use as baitfish for tuna in Puerto, Philippines. (Photo credit: Katherine Jack)

common, as are specifically designed hooks (squid jigs) used to catch squid. Diving with spear guns is common in the shallows, gleaning occurs along shorelines and beach seines are used in some locations. Many of the fish caught are locally processed and sold in local or regional markets, or in larger destinations such as Manila.

China is a particularly lucrative market for fishers in the Western Philippines from which demand has intensified in recent decades. Since the opening up of China's economy from 1978, incomes in China increased, as did demand for specific marine products, such as dried sea cucumbers, live reef food fish, shark fin and fish maw. The high prices paid for these marine products served as a catalyst for fishers throughout the Philippines to focus on them. The exploitation of sea cucumbers and live fish in Western Philippines highlights the dual phases of commodity frontiers, encompassing both commodity 'widening' with geographic expansion, and 'deepening' with the use of advanced technologies.

More than 30 types of sea cucumbers are exploited in the waters of the Philippines (Jontila et al., 2018). Once the sea cucumbers are caught they are dried and processed and sold through various market channels to eventually arrive in their destination markets, of which China is the largest. Consumption of sea cucumbers has been popular for centuries in China as a status and health food, and demand has spiked since the 1980s and 1990s (Eriksson et al., 2015). The most expensive tropical sea cucumbers are sandfish (*Holothuria scabra*), white teatfish (*Holothuria fuscogilva*), and black teatfish (*Holothuria whitmaei*), with prices reaching well over US$100 per kilogram in Chinese markets (Brown et al., 2010; Purcell et al., 2018). In their dried form (bêche-de-mer), sea cucumbers are shelf stable for weeks or months, so have offered a rare opportunity for remote coastal areas characterised by a lack of easy market access (Barclay et al., 2016).

Akamine's (2001) study of sea cucumber exploitation in Southern Palawan shows how fishing livelihoods responded to these market drivers. In the late 1970s, sea cucumbers were caught by skin divers, on relatively short trips closer to shore. While the use of air compressors was introduced around this time (i.e., using an air compressor on a boat and diving with a hose to breathe), many accidents occurred, and so this technology did not become popular until the arrival of more experienced fishers from the Visayan group of islands in the late 1980s. From around this time, vessels started to travel further into the South China Sea, diving deeper and targeting more types of species that became progressively more commercially valuable. Women, who previously participated in gleaning and inshore fishing for sea cucumbers, did not participate in these trips. From the 1990s, the depths to which divers would go to find sea cucumbers increased (up to 60 metres), and the use of depth sounders ('fishfinders') was also introduced. However, by the late 2000s, the sea cucumber fishery in Palawan (Brown et al., 2010) and in the Philippines more generally (Choo, 2008) had declined significantly. Overharvesting meant that the trade was characterised by a higher proportion of smaller sea cucumbers and of lower-valued species (Akamine, 2005; Brown et al., 2010), and local extirpations occurred. In Southern Palawan, while sea cucumbers continue to be harvested as a supplemental livelihood activity (Fabinyi et al., 2012), many fishers turned instead to another lucrative marine product: live reef fish.

Live reef fish have long served as an important component of seafood banquets in China, and, as with sea cucumbers, their demand has dramatically increased as wealth levels in the Chinese economy have grown since the 1980s. Particularly highly valued reef fish in these banquets include Napoleon wrasse (*Cheilinus undulatus*), humpback grouper (*Cromileptes altivelis*) and leopard coral grouper (*Plectropomus leopardus*). The fish are caught live and kept alive until they reach a restaurant. The vast majority of higher-valued reef fish end up in China. Exploitation of live reef fish expanded geographically over time, from waters near Hong Kong to the wider Indo-Pacific (Scales et al., 2006), including the Western Philippines. In Palawan, which supplies most of the country's live fish exports (Padilla et al., 2003), the live fish industry began in Coron in the late 1980s, and from there the trade spread throughout the municipalities of Palawan, all the way down to Balabac in the extreme South. While Coron remained an important trading hub, fewer fish came to be sourced from the waters around Coron due to overexploitation, and municipalities further south formed the epicentre of this trade in Palawan. While many attempts have been made to govern the trade in a more environmentally sustainable way in Palawan and elsewhere, institutions for sustainability have found it difficult to compete against the economic pressures of this lucrative fishery (Fabinyi & Dalabajan, 2011; Sadovy de Mitcheson et al., 2013).

In many instances, the expansion of the live reef fish fishery has been financed with capital originating from further up the commodity chain. Fishers are financed for their fishing trips (many of which last several days, or even beyond a week) by buyers based in the municipal towns. In many cases, these buyers are agents of exporters based in Manila, who in turn have financial relationships with importers based in Hong Kong. In the south of Palawan, many traders are financed by ethnic Chinese buyers (*towkays*) based in Sabah, Malaysia.

The live reef fish fishery is notable for its high dependence on particular technologies. While many fishers use weighted hooks with lures to catch live fish, the use of cyanide and air compressors is also common. Fishers dive deep with the assistance of a compressor and squirt a solution of cyanide into reefs to stun and catch the fish. If a fish is brought up too quickly from deep water, the swim bladder will rupture, killing it, and so fishers have to be skilled in puncturing the swim bladder with a hypodermic needle, obtained from local health clinics. Once the fish are brought to the surface, they are kept alive in specially designed aquariums in the vessels that allow fresh water to continuously flow through. If a fish is below optimal market size, it is placed in a grow-out cage for weeks or even months before being sold. Since the leopard coral groupers lose their bright red colour when staying for long periods of time in shallow water, many of these cages are located tens of metres below the surface. Fishers dive to feed them, using air compressors for breathing. After the fish are sold to a buyer's aquarium, the fish are often fed antibiotics and tranquilisers to reduce their mortality and stress during transport. The fish are then transported in oxygenated bags to the local airport, where they are flown to Manila, and subsequently transferred to a commercial flight to Hong Kong. The expansion of this trade in Palawan has, therefore, been highly dependent on the capital originating from buyers at higher levels of the commodity chain, and on the expansion of the physical infrastructure (roads, airports) and use of technologies (cyanide, needles, air compressors, medicines) required to catch and transport these fish.

Not only has the practice of fishing activities changed over time, as these examples of sea cucumbers and live fish show, but also the significance of fishing within the broader spectrum of activities that constitute household livelihoods in the coastal Philippines is dynamic. While specialist fishing communities that rely almost entirely on fishing as a livelihood remain common in the Philippines (Spoehr, 1984), especially in contexts where there are few other viable livelihood options, there are also many instances in which fishing is combined with other sources of income, such as farming, livestock raising, small household enterprises such as mixed-goods stores, and transport work (Eder, 2003). In these instances, fishing can be combined in a highly flexible manner, taken up in a seasonal, part-time or supplemental fashion. Fishers also typically practise multiple types of fishing activity at different times of the day, month or year.

As fishing activities in general become increasingly difficult in parts of the Western Philippines due to lower catches and increased pressure from regulations (e.g., MPAs), residents in some cases are turning to additional or alternative sources of income. For example, the growth of aquaculture in many regions of the Philippines has generated opportunities for fishers. In the Western Philippines the government has done much to stimulate seaweed production through support programmes. While this activity contributes to livelihood portfolios, in many cases as a supplemental livelihood activity, without substantial investments it rarely generates the sorts of profits found in the sea cucumber and live reef food fisheries. In parts of coastal Mindoro, which has had a longer history of settlement and economic diversification than Palawan, remittances from family members working overseas also now form a considerable proportion of household incomes.

In particular, fishing livelihoods in the Western Philippines have been adjusting to the rapid rise in tourism. Promoted heavily by governments in the Philippines at all levels, tourism is widely viewed as an economic activity that can generate economic benefits and to do so in a more environmentally sustainable manner than many fisheries. Not every community in Western Palawan is regularly frequented by tourists, and a range of positive and negative effects of coastal tourism has been identified (Fabinyi, 2020). However, the growth in recent years has been enough to drive many previous full-time fishers into livelihoods based instead on tourism (e.g., guesthouse accommodation, converting fishing boats into 'beach-hopping' or dive boats, guiding, etc.), or mixing tourism and fishing livelihood activities (e.g., supplying restaurants with seafood). The growth in related infrastructure (airports, roads, buildings) has also drawn people to work in construction and other wage labour jobs. In 2020, the COVID-19 pandemic effectively shut down the tourism industry in Western Philippines, so many people again turned to fishing as their main livelihood.

Papua New Guinea

In contrast to the Philippines, PNG is a country with a much lower population density. According to the World Bank, in 2019 the population of the Philippines was around 108 million for 300,000 km², whereas the PNG population was around 8.7 million for 463,000 km². In general there are much lower levels of industrialisation in PNG, and subsistence horticulture remains the most important livelihood. Fishing livelihoods are in general much less diverse, with most fishing being for home consumption or sale in local markets, resulting in significantly less fishing pressure on the marine environment. Yet, both during colonial times and more recently, coastal zones of PNG have also been rapidly drawn into trading networks at multiple scales, and these trade networks and market demands and opportunities exercise significant influence over the sorts of fishing livelihoods available to coastal residents.

In pre-colonial times, internal migration around PNG was relatively limited in geographic extent. After European contact and settlement, there was a reduction of movement in response to violence and warfare, but over time more people began to migrate to government and mission stations, and urban centres (May & Skeldon, 1977). This rural-to-urban migration has meant that in some coastal communities where fishing is a major livelihood, remittances from urban centres (e.g., Port Moresby) are significant—although this is variable (Carrier, 1981; Hayes, 1993; Vieira et al., 2017). In recent decades, in addition to well-established forms of rural-to-urban migration, internal migration has often been characterised by people moving towards large-scale resource extraction projects such as mines (Bainton, 2017) or agricultural plantations (Curry & Koczberski, 1998). Unlike in the Philippines, fishing is rarely the main motivation for migration in coastal villages. In a study of 14 coastal villages across PNG, not one migrant respondent nominated fishing as their reason for migration, most nominating other employment opportunities and marriage as their reasons (Cinner, 2009). However, thousands of internal migrants living in settlements around the cities of Madang and Lae work in the tuna industries based around those cities (Barclay, 2012).

Nevertheless, the ways in which migration patterns affect livelihoods and resource use in coastal communities remain highly important, largely due to the social relationships between migrants and non-migrants. In terrestrial parts of PNG, Filer (1997) has documented the emergence of an 'ideology of landownership', where the growth of resource extraction projects from the 1980s led to heightened consciousness of and identification with customary landownership. This has led, for example, to contestations and disputes over who is a rightful landowner, to whom one has social obligations, and who should, therefore, benefit from the activities of mines (Bainton, 2009).

Fishing and marine tenure, similar to terrestrial sectors, have also been subject to disputes about rightful ownership (e.g., Kinch, 2020; see also Foale & Macintyre, 2000 for Solomon Islands). Central to these disputes are ideas about who holds customary rights. In his study of coastal communities across PNG, Cinner (2009) noted that migrants were in many cases excluded from access to marine resources, and were less involved in decision-making. Therefore, while coastal fisheries are not a pull factor for migration to coastal areas in PNG, patterns of migration do inform who is able to fish as a livelihood. As Connell and Lutkehaus (2017: 92) note in their study of resettlement projects in coastal PNG, 'social relationships are written in the ground and editing or removing the writing is almost impossible, establishing a geometry of power that absolutely marginalises potential settlers'.

Central to the increasing emphasis on access rights to marine resources has been engagement with markets. Exchange practices in coastal zones of PNG have been a crucial part of life from pre-colonial times (Malinowski, 1922). With the emergence of marketplaces in urban areas during the colonial period (Busse & Sharp, 2019), fish have been sold for cash income. In a study of six sites in coastal PNG, Cinner and McClanahan (2006: 78) found that 'more than half of the caught fish were bartered or sold'. Similarly, in a study from Madang Province, Havice and Reed (2012: 424) note that fish catch is transitioning from consumption to selling in markets for cash, while in Manus, Lau et al. (2020) found that selling fish in the market was the preferred use for fish, over bartering, sharing or eating. Most urban areas and many coastal rural areas in PNG have marketplaces that sell diverse species of fish caught from a range of gears, raw and cooked (Busse & Sharp, 2019; Cinner & McClanahan, 2006). Specific technologies have emerged together with the expansion of domestic markets that have also increased the capacity of people to fish further distances, use new techniques for catching fish and to access further markets. These include the use of outboard motors, fibreglass boats, synthetic lines, metal hooks, compressors for diving, ice and ice chests and fish aggregating devices.

In addition to the development of catching fish for local marketplaces, a variety of export fisheries has emerged in PNG, including tuna (Barclay & Cartwright, 2008; Havice & Reed, 2012), aquarium fish (Máñez et al., 2014), live reef food fish (Hamilton & Matawai, 2006) and dried sea products. In coastal areas, bêche-de-mer and other dried products such as shark fin have been traded to South-East Asia since the 1800s, but became a significant industry in PNG from the early 1990s, as demand for these products in China boomed (Barclay et al., 2019; Kinch, 2020). Sea cucumbers are harvested usually by groups, dried and processed locally, and then transported to provincial capitals for trade onwards (see Fig. 2.2). In the

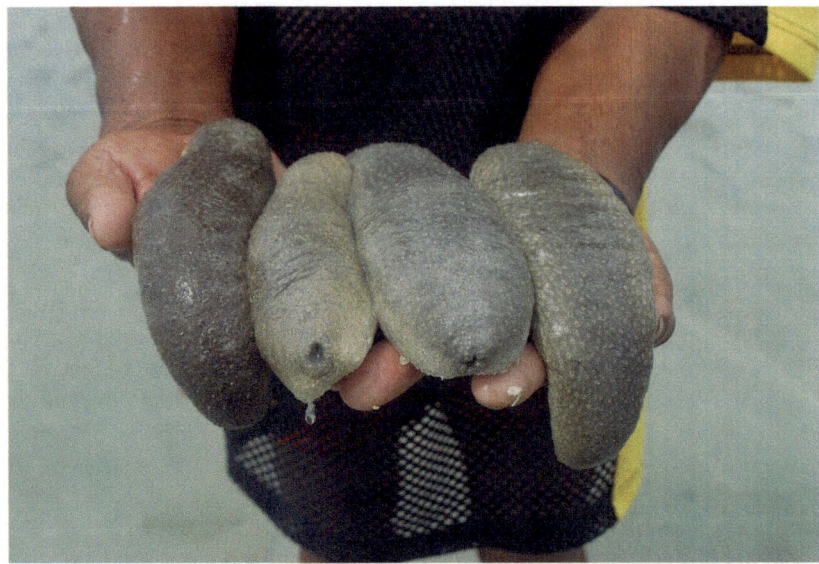

Fig. 2.2 Fisher holding freshly caught sea cucumbers. (Photo credit: Arselene Uyami-Bitara)

1990s the bulk of the trade went to Singapore and Malaysia, but from the mid-2000s Hong Kong and China became the main destinations.

In many coastal parts of Milne Bay Province, as with other coastal areas in PNG, sales of bêche-de-mer became the most important source of cash income, almost exclusively in some places (Foale, 2005; Kinch, 2020). The strong demand for bêche-de-mer translated into high prices that dwarfed other income-generating opportunities, so fishing livelihoods became largely focused on this one commodity. While this increase in cash income led to benefits for many families, including basic necessities such as food, it also generated social challenges. As much of the diving for sea cucumbers was done by physically capable young men, they subsequently ended up controlling much of the cash, with tensions among younger and older men (Rasmussen, 2015), and between men and women (Barclay et al., 2016; Barclay et al., 2019). The rapid increase in the value of marine resources also led to protracted disputes among groups over access to fishing grounds (Foale, 2005; Kinch, 2020).

The consequences of this intensified effort were that from the mid-2000s sea cucumber stocks declined precipitously. As in the Philippines, PNG fishers shifted their attention from higher-value species to lower-value species, meaning they had to take even greater amounts to maintain incomes (Barclay et al., 2019). In 2009, a moratorium was instituted to ban the sale and trade of bêche-de-mer.[2] With the sudden cessation of income from sea cucumber, fishers were forced to shift into other livelihood activities. The amount of cash income derived by many coastal communities declined significantly (Barclay et al., 2019; Vieira et al., 2017).

While livelihood activities have always been mixed in coastal PNG, subsistence gardening remains a core component of most coastal livelihood portfolios. For example, sweet potatoes, bananas and taro are grown by 99 per cent, 96 per cent and 95 per cent of the PNG population, respectively (Bourke & Allen, 2009: 195). In coastal areas such as Milne Bay, while some income-generating opportunities remained after the bêche-de-mer ban was imposed, such as in copra plantations, overall, livelihoods became focused again on gardening. However, in some places gardening productivity was reduced by years of neglect, as people had focused on bêche-de-mer fishing (Barclay et al., 2016: 39). Although shark fin and trochus shell remain relatively important as cash-earning commodities in Milne Bay (Vieira et al., 2017), the fishing component of livelihoods has reverted to being more of a supplemental activity generating food and some cash income. Thus, as climate change effects increase (Connell & Lutkehaus, 2017) and the stocks of vulnerable, high-value species such as sharks and sea cucumbers decline, fishing livelihoods in PNG will continue to evolve in relation to the opportunities afforded by migration and other land-based livelihood activities, in particular farming.

CONCLUSION

While fishing livelihoods have been practised for millennia, they are not static. Even in economically remote parts of the Asia-Pacific, fishers have responded to market demands from nearby and beyond. These market demands shape what kinds of fish are targeted, what technologies are used in the catch, processing and distribution of fish, and how fishing activities relate to other livelihood activities, many of which are similarly shaped by other market demands. In many cases, these market demands and

[2] This moratorium was lifted in 2017.

opportunities are also a major factor behind where people choose to live. Ultimately, changing markets and population densities strongly affect the status of fisheries and the conditions for their sustainability (Cinner et al., 2013).

The interactions between fishing livelihoods and these broader global forces are mediated by very different contexts in PNG and Philippines. Cultures (Chap. 3) and governance (Chap. 4) are very different in these countries, and the Philippines has a significantly greater degree of economic integration with local and international markets, and a much larger, more densely distributed population than PNG. Yet, despite these different contexts, both countries have experienced increases in the geographic scale and the technological intensity of fishing activities—commodity 'widening' and 'deepening' (Moore, 2015). Taking into consideration historical trajectories and market drivers, we can see how fishing livelihoods are influenced not only by individual or household decision-making, or by local or national governance structures, but also by wider, systemic forces of global economic transformation.

In many cases, these wider forces of global capitalism and development have favoured fishing activities to the point that they have become biologically unsustainable. This has flow-on effects for fishers who have to adapt to target other fish, or adopt new livelihood activities beyond capture fisheries alone. The extent to which fishing livelihoods integrate, compete with or are ultimately surpassed by newer forms of coastal livelihoods such as tourism and aquaculture will be a major part of fishing livelihoods in the future.

For fisheries managers and policymakers, understanding the historical trajectories of fishing livelihoods, how they have changed and adapted over time, and how they are integrated with the wider economy provides important context on the external drivers of fishing activity and how fishers are likely to behave. For example, the relationship of fishing livelihoods to economic activities in other sectors is important when trying to generate 'alternative' livelihoods and encourage fishers to exit from the fishery, or when implementing regulations that rely on some degree of reduced fishing effort (Barclay et al., 2019). While economic and market-based approaches to fisheries governance attempt to work with individual markets, this approach can potentially conceal the wider systemic forces at play—the logics of commodity widening and deepening that ultimately drive further exploitation.

REFERENCES

Akamine, J. (2001). Holothurian exploitation in the Philippines: Continuities and discontinuities. *Tropics, 10*(4), 591–607. https://doi.org/10.3759/tropics.10.591

Akamine, J. (2005). Role of the trepang traders in the depleting resource management: A Philippine case. In N. Kishigami & J. M. Savelle (Eds.), *Indigenous use and management of marine resources* (pp. 259–278). National Museum of Ethnology.

Akram-Lodhi, A. H., & Kay, C. (2010). Surveying the Agrarian question (part 2): Current debates and beyond. *The Journal of Peasant Studies, 37*(2), 255–284. https://doi.org/10.1080/03066151003594906

Baglioni, E., & Campling, L. (2017). Natural resource industries as global value chains: Frontiers, fetishism, labour and the state. *Environment and Planning A: Economy and Space, 49*(11), 2437–2456. https://doi.org/10.1177/0308518X17728517

Bainton, N. A. (2009). Keeping the network out of view: Mining, distinctions and exclusion in Melanesia. *Oceania, 79*(1), 18–33. https://doi.org/10.1002/j.1834-4461.2009.tb00048.x

Bainton, N. A. (2017). Migrants, labourers and landowners at the Lihir Gold Mine, Papua New Guinea. In C. Filer & P.-Y. Le Meur (Eds.), *Large-scale mines and local-level politics* (pp. 313–351). ANU Press.

Barclay, K. (2012). Social impacts. In Blomeyer & Sanz (Ed.), *Application of the system of derogation to the rules of origin of fisheries products in Papua New Guinea and Fiji* (pp. 155–188). European Parliament.

Barclay, K. (2014). History of industrial tuna fishing in the Pacific Islands. In J. Christensen & M. Tull (Eds.), *Historical perspectives of fisheries exploitation in the Indo-Pacific* (pp. 153–171). MARE Series Vol. 12. Springer.

Barclay, K., & Cartwright, I. (2008). *Capturing the wealth from tuna: Case studies from the Pacific.* ANU Press.

Barclay, K., Kinch, J., Fabinyi, M., Waddell, S., Smith, G. K., Sharma, S., Hamilton, R., et al. (2016). *Interactive governance analysis of the bêche-de-mer 'fish chain' from Papua New Guinea to Asian markets.* University of Technology Sydney. https://doi.org/10.13140/RG.2.2.10787.66083

Barclay, K., Fabinyi, M., Kinch, J., & Foale, S. (2019). Governability of high-value fisheries in low-income contexts: A case study of the sea cucumber fishery in Papua New Guinea. *Human Ecology, 47*, 381–396. https://doi.org/10.1007/s10745-019-00078-8

Belton, B., & Thilsted, S. H. (2014). Fisheries in transition: Food and nutrition security implications for the global South. *Global Food Security, 3*(1), 59–66.

Béné, C., Steel, E., Luadia, B. K., & Gordon, A. (2009). Fish as the 'bank in the water'—Evidence from chronic-poor communities in Congo. *Food Policy, 34*(1), 108–118.

Bernstein, H. (2010). *Class dynamics of agrarian change* (Vol. 1). Kumarian Press.

Bourke, R. M., & Allen, B. (2009). Village food production systems. In R. M. Bourke & T. Harwood (Eds.), *Food and agriculture in Papua New Guinea* (pp. 193–269). ANU Press.

Brown, E. O., Perez, M. L., Garces, L. R., Ragaza, R. J., Bassig, R. A., & Zaragoza, E. C. (2010). *Value chain analysis for sea cucumber in the Philippines.* WorldFish Center Studies and Reviews No. 2120. The WorldFish Center.

Busse, M., & Sharp, T. L. M. (2019). Marketplaces and morality in Papua New Guinea: Place, personhood and exchange. *Oceania, 89*(2), 126–153. https://doi.org/10.1002/ocea.5218

Butcher, J. G. (2004). *The closing of the frontier: A history of the marine fisheries of Southeast Asia, c. 1850–2000.* Institute of Southeast Asian Studies.

Campling, L. (2012). The tuna 'commodity frontier': Business strategies and environment in the industrial tuna fisheries of the Western Indian Ocean. *Journal of Agrarian Change, 12*(2–3), 252–278. https://doi.org/10.1111/j.1471-0366.2011.00354.x

Campling, L., & Havice, E. (2018). The global environmental politics and political economy of seafood systems. *Global Environmental Politics, 18*(2), 72–92. https://doi.org/10.1162/glep_a_00453

Carrier, J. G. (1981). Ownership of productive resources on Ponam Island, Manus province. *Journal de la Societe des Oceanistes, 37*(72), 205–217. https://doi.org/10.3406/jso.1981.3061

Chen, H. (2008a). Japan and the birth of Takao's fisheries in Nanyo, 1895–1945. *International Journal of Maritime History, 20*(1), 133–152. https://doi.org/10.1177/084387140802000107

Chen, T.-Y. (2008b). The involvement of fishers of Xiao Liuqiu in the Southeast Asia tuna fishery, 1945–1980. *MAST Journal of Maritime Studies, 2*(2), 51–74. Retrieved February 5, 2021, from http://www.marecentre.nl/mast/documents/Mast_7_1_Chen.pdf

Choo, P.-S. (2008). The Philippines: A hotspot of sea cucumber fisheries in Asia. In V. Toral-Granda, A. Lovatelli, & M. Vasconcellos (Eds.), *Sea cucumbers. A global review of fisheries and trade* (pp. 119–140). FAO Fisheries and Aquaculture Technical Paper No. 516. FAO.

Christensen, J. (2014). Unsettled seas: Towards a history of marine animal populations in the Central Indo-Pacific. In J. Christensen & M. Tull (Eds.), *Historical perspectives of fisheries exploitation in the Indo-Pacific* (pp. 13–39). MARE Publication Series 12. Springer.

Cinner, J. E. (2009). Migration and coastal resource use in Papua New Guinea. *Ocean & Coastal Management, 52*(8), 411–416. https://doi.org/10.1016/j.ocecoaman.2009.06.003

Cinner, J. E., & McClanahan, T. R. (2006). Socioeconomic factors that lead to overfishing in small-scale coral reef fisheries of Papua New Guinea. *Environmental Conservation, 33*(1), 73–80. https://doi.org/10.1017/S0376892906002748

Cinner, J. E., Graham, N. A. J., Huchery, C., & MacNeil, M. A. (2013). Global effects of local human population density and distance to markets on the condition of coral reef fisheries. *Conservation Biology, 27*(3), 453–458. https://doi. org/10.1111/j.1523-1739.2012.01933.x

Connell, J., & Lutkehaus, N. (2017). Environmental refugees? A tale of two resettlement projects in coastal Papua New Guinea. *Australian Geographer, 48*(1), 79–95. https://doi.org/10.1080/00049182.2016.1267603

Curry, G., & Koczberski, G. (1998). Migration and circulation as a way of life for the Wosera Abelam of Papua New Guinea. *Asia Pacific Viewpoint, 39*(1), 29–52.

Eder, J. F. (2003). Of fishers and farmers: ethnicity and resource use in coastal Palawan. *Philippine Quarterly of Culture and Society, 31*(3), 207–225.

Eder, J. F. (2008). *Migrants to the coasts: Livelihood, resource management, and global change in the Philippines.* Nelson Education.

Eder, J. F., & Fernandez, J. O. (Eds.). (1996). *Palawan at the crossroads: Development and the environment on a Philippine frontier.* Ateneo de Manila University Press.

Eriksson, H., Österblom, H., Crona, B., Troell, M., Andrew, N., Wilen, J., & Folke, C. (2015). Contagious exploitation of marine resources. *Frontiers in Ecology and the Environment, 13*(8), 435–440. https://doi. org/10.1890/140312

Fabinyi, M. (2020). The role of land tenure in livelihood transitions from fishing to tourism. *Maritime Studies, 19*(1), 29–39. https://doi.org/10.1007/s40152-019-00145-2

Fabinyi, M., & Dalabajan, D. (2011). Policy and practice in the live reef fish for food trade: A case study from Palawan, Philippines. *Marine Policy, 35*(3), 371–378. https://doi.org/10.1016/j.marpol.2010.11.001

Fabinyi, M., Pido, M., Harani, B., Caceres, J., Uyami-Bitara, A., & De las Alas, A., Ponce de Leon, E. M., et al. (2012). Luxury seafood consumption in China and the intensification of coastal livelihoods in Southeast Asia: The live reef fish for food trade in Balabac, Philippines. *Asia Pacific Viewpoint, 53*(2), 118–132. https://doi.org/10.1111/j.1467-8373.2012.01483.x

Filer, C. (1997). Compensation, rent and power in Papua New Guinea. In S. Toft (Ed.), *Compensation for resource development in Papua New Guinea* (pp. 156–189). Port Moresby: Law Reform Commission of Papua New Guinea; Resource Management in Asia and the Pacific; Research School of Pacific and Asian Studies and National Centre for Development Studies, Australian National University.

Firth, R. (1966). *Malay fishermen: Their peasant economy* (2nd ed.). Routledge and Kegan Paul.

Foale, S. (2005). *Sharks, sea slugs and skirmishes: Managing marine and agricultural resources on small, overpopulated islands in Milne Bay, PNG.* Working Paper No. 64. Canberra: Resource Management in Asia-Pacific Program, Division of Pacific and Asian History, and Research School for Pacific and Asian Studies, Australian National University.

Foale, S., & Macintyre, M. (2000). Dynamic and flexible aspects of land and marine tenure at West Nggela: Implications for marine resource management. *Oceania*, *71*(1), 30–45. https://doi.org/10.1002/j.1834-4461.2000.tb02722.x

Fujinami, N. (1987). Development of Japan's tuna fisheries. In D. J. Doulman (Ed.), *Tuna issues and perspectives in the Pacific Islands region* (pp. 57–70). East-West Center.

Hamilton, R. J., & Matawai, M. (2006). Live reef food fish trade causes rapid declines in abundance of squaretail coralgrouper (*Plectropomus areolatus*) at a spawning aggregation site in Manus, Papua New Guinea. *SPC Live Reef Fish Information Bulletin*, *16*, 13–18.

Harvey, D. (1982). *The limits to capital*. Blackwell.

Harvey, D. (1989). *The condition of postmodernity* (Vol. 14). Blackwell.

Havice, E., & Reed, K. (2012). Fishing for development? Tuna resource access and industrial change in Papua New Guinea. *Journal of Agrarian Change*, *12*(2–3), 413–435. https://doi.org/10.1111/j.1471-0366.2011.00351.x

Hayes, G. (1993). 'MIRAB' processes and development on Small Pacific Islands: A case study from the Southern Massim, Papua New Guinea. *Pacific Viewpoint*, *34*(2), 153–177. https://doi.org/10.1111/apv.342002

Helbling, J., & Schult, V. (1997). Demographic development in Mindoro. *Philippine Studies*, *45*(3), 385–407.

Jontila, J. B. S., Monteclaro, H. M., Quinitio, G. F., Santander-de Leon, S. M., & Altamirano, J. P. (2018). The sea cucumber fishery in Palawan, Philippines. *Kuroshio Science*, *12*(1), 84–88.

Kinch, J. (2020). *Changing lives and livelihoods: Culture, capitalism and contestation over marine resources in Island Melanesia*. Doctoral dissertation. Canberra: Australian National University. https://doi.org/10.25911/5e9ecc08d7068

Lau, J. D., Cinner, J. E., Fabinyi, M., Gurney, G. G., & Hicks, C. C. (2020). Access to marine ecosystem services: Examining entanglement and legitimacy in customary institutions. *World Development*, *126*, 104–730. https://doi.org/10.1016/j.worlddev.2019.104730

Malinowski, B. (1922). *Argonauts of the Pacific*. Routledge.

Máñez, K. S., Dandava, L., & Ekau, W. (2014). Fishing the last frontier: The introduction of the marine aquarium trade and its impact on local fishing communities in Papua New Guinea. *Marine Policy*, *44*, 279–286. https://doi.org/10.1016/j.marpol.2013.09.018

May, R. J., & Skeldon, R. (1977). Internal migration in Papua New Guinea: An introduction to its description and analysis. In R. J. May (Ed.), *Change and movement: Readings on internal migration in Papua New Guinea* (pp. 1–26). Papua New Guinea Institute of Applied Social and Economic Research in association with Australian National University Press.

Minnegal, M., King, T. J., Just, R., & Dwyer, P. D. (2003). Deep identity, shallow time: Sustaining a future in Victorian fishing communities. *The Australian Journal of Anthropology, 14*(1), 53–71. https://doi.org/10.1111/j.1835-9310.2003.tb00220.x

Moore, J. (2015). *Capitalism in the web of life: Ecology and the accumulation of capital*. Verso.

Morgan, G. R., & Staples, D. J. (2006). Tuna longlining, poling and purse seining. In G. R. Morgan, & D. J. Staples (Eds.), *The history of industrial marine fisheries in Southeast Asia*. UN FAO Regional Office for Asia and the Pacific. Retrieved February 5, 2021, from http://www.fao.org/3/AG122E00.htm#Contents

Padilla, J. E., Mamauag, S., Braganza, G., Brucal, N., Yu, D., & Morales, A. (2003). *Sustainability assessment of the live reef-fish for food industry in Palawan Philippines*. Philippines: World Wildlife Fund.

Pálsson, G., & Durrenberger, E. P. (1990). Systems of production and social discourse: The skipper effect revisited. *American Anthropologist, 92*(1), 130–141. https://doi.org/10.1525/aa.1990.92.1.02a00090

Platteau, J.-P. (1984). The drive towards mechanization of smallscale fisheries in Kerala: A study of the transformation process of traditional village societies. *Development and Change, 15*(1), 65–103. https://doi.org/10.1111/j.1467-7660.1984.tb00174.x

Purcell, S. W., Mercier, A., Conand, C., Hamel, J. F., Toral-Granda, M. V., Lovatelli, A., & Uthicke, S. (2013). Sea cucumber fisheries: Global analysis of stocks, management measures and drivers of overfishing. *Fish and Fisheries, 14*(1), 34–59. https://doi.org/10.1111/j.1467-2979.2011.00443.x

Purcell, S. W., Williamson, D. H., & Ngaluafe, P. (2018). Chinese market prices of bêche-de-mer: implications for fisheries and aquaculture. *Marine Policy, 91*, 58–65. https://doi.org/10.1016/j.marpol.2018.02.005

Ransan-Cooper, H. (2016). The role of human agency in environmental change and mobility: A case study of environmental migration in Southeast Philippines. *Environmental Sociology, 2*(2), 132–143. https://doi.org/10.1080/2325104 2.2016.1144405

Rasmussen, A. E. (2015). *In the absence of the gift: New forms of value and personhood in a Papua New Guinea community*. Berghahn Books.

Robins, L., Crimp, S., van Wensveen, M., Alders, R. G., Bourke, R. M., Butler, J., Cosijn, M., et al. (2020). *COVID-19 and food systems in the Indo-Pacific: An assessment of vulnerabilities, impacts and opportunities for action*. ACIAR Technical Report No. 96. Canberra: ACIAR.

Sadovy de Mitcheson, Y., Craig, M. T., Bertoncini, A. A., Carpenter, K. E., Cheung, W. W., Choat, J. H., Liu, M., et al. (2013). Fishing groupers towards extinction: A global assessment of threats and extinction risks in a billion dollar fishery. *Fish and Fisheries, 14*(2), 119–136. https://doi.org/10.1111/j.1467-2979.2011.00455.x

Sadovy de Mitcheson, Y., To, A.W.-L., Wong, N. W., Kwan, H. Y., & Bud, W. S. (2019). Emerging from the murk: Threats, challenges and opportunities for the global swim bladder trade. *Reviews in Fish Biology and Fisheries*, 1–27. https://doi.org/10.1007/s11160-019-09585-9

Scales, H., Balmford, A., Liu, M., Sadovy, Y., Manica, A., Hughes, T. P., Berkes, F., et al. (2006). Keeping bandits at bay? *Science, 313*(5787), 612–614. https://doi.org/10.1126/science.313.5787.612c

Schult, V. (1991). The genesis of lowland Filipino society in Mindoro. *Philippine Studies, 39*(1), 92–103.

Seki, K. (2004). Maritime migration in the Visayas: A case study of the Dalaguetenon fisherfolk in Cebu. In H. Umehara & G. M. Bautista (Eds.), *Communities at the margins: Reflections on social, economic and environmental change in the Philippines* (pp. 193–221). Ateneo de Manila University Press.

Spoehr, A. (1984). Change in Philippine capture fisheries: An historical overview. *Philippine Quarterly of Culture and Society, 12*(1), 25–56.

Swartz, W., Sumaila, U. R., Watson, R., & Pauly, D. (2010). Sourcing seafood for the three major markets: The EU, Japan and the USA. *Marine Policy, 34*(6), 1366–1373.

Ushijima, I., & Zayas, C. N. (Eds.). (1994). *Fishers of the Visayas*. College of Social Sciences and Philosophy, University of the Philippines.

Vieira, S., Kinch, J., White, W., & Yaman, L. (2017). Artisanal shark fishing in the Louisiade Archipelago, Papua New Guinea: socio-economic characteristics and management options. *Ocean & Coastal Management, 137*, 43–56. https://doi.org/10.1016/j.ocecoaman.2016.12.009

Wallerstein, I. (2004). *World-systems analysis*. Duke University Press.

CHAPTER 3

Fishing Livelihoods and Social Diversity

Abstract This chapter shifts scale from Chap. 2 to focus on the local context and analyse the everyday sets of social relationships that frame the lives of those engaged in fishing livelihoods. The broad structural forces of migration, technology and markets along with the wider economy all intersect with local sets of social structures to shape the conditions in which fishing livelihoods operate. Here we present two examples of how different forms of social differentiation interact with fishing livelihoods. In the Western Philippines, class and status intersect with cultural values to generate power relations and hierarchies in different roles associated with fishing livelihoods. In Pacific Island countries, gender norms structure the different types of fishing activities in which men and women are involved.

Keywords Class • Status • Gender • Philippines • Oceania • Value chain

This chapter shifts scale from Chap. 2 to focus on the local context and analyse the everyday sets of social relationships that frame the lives of those engaged in fishing livelihoods. The broad structural forces of migration, technology and markets along with the wider economy all intersect with local sets of social structures to shape the conditions in which fishing livelihoods operate. Understanding how these forms of social relationships—such as class, gender and ethnicity—operate in relation to fishing

© The Author(s) 2022
M. Fabinyi, K. Barclay, *Asia-Pacific Fishing Livelihoods*,
https://doi.org/10.1007/978-3-030-79591-7_3

livelihoods matters, because it shows how both fishing livelihoods and governance projects to manage these livelihoods are socially differentiated.

Much policy-oriented literature and practice in fisheries takes as relatively unproblematic starting points the ideas of a 'community' and a 'fisher'. Yet, an individual is far more than a 'fisher' whose sole priority is to catch fish, and the idea of a 'fishing community' disguises a range of social cleavages, hierarchies and identities within groups. Different people will have different levels of engagement in fishing, different types of roles within fishing, and different expectations and understandings about fishing—all of which affect how we understand what their particular version of a fishing livelihood is, and how governors seek to manage it. For example, dominant narratives about fishers and poverty in developing countries (e.g., 'fishers are poor because they fish'; Béné, 2003) can lead to governance interventions that ignore the wider context of vulnerability in which fishers may live and that they prioritise (e.g., lack of access to health care, lack of land tenure or inequalities among different social groups) (Béné & Friend, 2011; Fabinyi et al., 2015; Mills et al., 2011). Without careful attention to social differentiation, new governance institutions for sustainability are liable to get 'sucked up' into these existing patterns of inequality across class, gender, ethnic and other lines (Eder, 2005).

Many studies of social differentiation take the concept of class as their starting point. Early literature in political ecology, and much literature in discussions of agrarian change, uses the concept of class as a key marker of social differentiation, analysing the diverse ways in which groups of people engage with markets and relate to the means of production (e.g., as worker or owner). As Bernstein (2010: 22) summarises, such an approach is largely informed by asking basic questions on resource use, ownership and distribution.[1]

In the large-scale or industrial fisheries sector, there are significant class distinctions between boat owners and crew, reflected in systems of profit sharing (McCall Howard, 2012). However, in many small-scale fisheries, distinctions between owners and crew are frequently much less distinct and can be thought of instead as a form of 'petty commodity production', where owners occupy dual roles of both capital and labour (Russell & Poopetch, 1990). Owners often work on their own vessels, employ crew through kin networks and have more egalitarian profit-sharing systems. In

---

[1] Specifically, 'Who owns what? Who does what? Who gets what? What do they do with it?'

many other types of small-scale fisheries, there is no paid crew at all (i.e., it may just be a fisher, or a fisher with a family member), and the catch may be only partially marketed or used entirely for consumption. There are also often sharp class distinctions along the value chain, from producer to trader to consumer.

Beyond hierarchical categorisations of class, income and wealth, there are other distinctions between different types of fishers. For example, younger fishers, who in some cases are undertaking illegal fishing practices (Fabinyi, 2012; Lowe, 2002), may have different sets of economic goals compared to older fishers. Full-time fishers who rely on fishing as a sole source of income have different perceptions about fishing compared to part-time or seasonal fishers. The extraordinary diversity of gears and techniques mentioned in Chap. 2 means that new forms of governance, such as MPAs, have highly differentiated effects for different types of fishers (Eder, 2005).

Yet, livelihood roles are not the only factors underlying social differentiation (Hornborg et al., 2013).[2] As political ecologists shifted from a primary focus on the forces of capitalism, they engaged with other ways of categorising social differentiation, including gender (Rocheleau et al., 1996), culture and ethnicity (Peet & Watts, 1996). In line with developments in social anthropology, this approach foregrounded the roles of meaning, identity and interpretation over a 'materialist' emphasis on 'protein and profit' (Geertz, 1973; Sahlins, 1978). Individuals have multiple identities that can shift over time and according to circumstance, or can intersect. Importantly, forms of social differentiation are not necessarily 'natural', but have elements of social construction and can be used in strategic ways (Dressler & Turner, 2008; Li, 2000). Ultimately, these markers of difference serve as crucial determinants of access to or exclusion from resources at the local level (Hall et al., 2011; Ribot & Peluso, 2003).

In many cases forms of social differentiation can map on to dominant or subordinate roles within fishing livelihoods or fisheries value chains. For example, women tend to be more closely associated with near-shore fishing and gleaning as well as onshore roles such as processing and trading (Weeratunge et al., 2010). In South-East Asia, the ethnic Sama-Bajau tend to follow small-scale fishing livelihoods (Stacey et al., 2018), while in

---

[2] As Hornborg et al. (2013: 2) note, 'an attribution of decisive significance to material parameters in reproducing power structures should not imply down-playing the role of socio-cultural categories in organizing such structures'.

South Asia different castes are associated with particular occupations, including fishing (Coulthard, 2008).

Here we present two examples of how different forms of social differentiation interact with fishing livelihoods. In the Western Philippines, class and status intersect with cultural values to generate power relations and hierarchies in different roles associated with fishing livelihoods. In Pacific Island countries, gender norms structure the different types of fishing activities in which men and women are involved.

Class and Status in Western Philippines Fishing Communities

Much social science of the political and economic development of the Philippines has highlighted relations of power and hierarchy (Kerkvliet, 1990; Sidel, 1999). In particular, inland agrarian regions, such as the sugar industry of Negros Island, or the rice-growing regions of Central Luzon, were characterised by sharp distinctions in assets and income between landowners and tenants, subsequently serving as the site of ongoing struggles over land reform. Compared to these agrarian regions, it can be more complex to characterise the forms of social differentiation in coastal communities (Eder, 2008). Yet, here too, economic roles and social identities interact to produce social institutions that condition forms of access and exclusion.

The classic division between the owner of the means of production and the wage labourer is reflected in various ways in coastal communities in the Western Philippines. In large-scale fishing vessels, owners of the vessels are typically located in urban spaces (e.g., municipal towns, provincial capital cities), and in many cases are owned by firms that may own several vessels. In these large-scale fisheries (e.g., lift nets, baby purse seines, trawlers) the distinction between boat owners and boat crew is, therefore, quite distinct, with the owners hiring the captain and crew. However, in rural areas of the Western Philippines where many small-scale fishers operate, divisions between people based on their relationship to the means of production can be less obvious. In coastal communities there are typically many diverse modes of fishing, using different gears to target different marine resources at different locations, at different times of the day, month and year, and individuals typically move across several fisheries at any one time.

Many individuals or households straddle the distinctions between labour and capital as 'petty commodity producers'. For example, a fisher may own a motorised boat and work on it independently while employing other people as crew. In these circumstances, crews are often recruited through kin and/or neighbour networks, and the distribution of profits is frequently through a share system. In this system of profit sharing, shares may be allocated for the owner of the vessel and/or gear, and then distributed to crew members based on either total or individual catch. Some of the common types of small-scale fisheries operating in municipal waters are based on small, motorised boats with around two–four crew, which use gillnets to catch small pelagic fish or seagrass-dwelling species such as rabbitfish (*Siganidae*), or use hook and line to target live reef fish (see Fig. 3.1). Some vessels operate a hybrid system between the small-scale, petty commodity mode of production and that used by industrial ventures. For example, vessels fishing for fusiliers, sardines or mixed reef fish may have more than 20 crew and go out for two or more weeks at a time, but the owner works on board as the captain and recruits crew through kin networks (Fabinyi, 2012).

Owners of motorised boats tend to be more visibly well-off than those who do not own a vessel. For those without capital in the form of a fishing vessel or gear, all of their fishing income must derive directly from how much they catch, and they tend to be among the poorest in any rural coastal community. Their housing is often made of temporary bush materials (*nipa*), and many have no access to electricity and go through periods of food insecurity (Fabinyi et al., 2017).

Frequently, the distinctions between boat owners and those without boats overlap with other forms of differentiation. Indigenous groups, such as the Mangyan in Mindoro and Tagbanua in Northern Palawan, tend to live inland or to participate in fisheries as hired crew (Dressler & Fabinyi, 2011). In particular, a key axis of differentiation in many coastal communities in the Western Philippines (as elsewhere in the country) relates to migrant status (Knudsen, 2012, 2016). More recent migrants tend to have limited assets and need to find work on boats through actively setting up social relationships with boat owners. More recent migrants tend to be more socially and economically marginalised than residents who migrated earlier and have established themselves.

Differences in wealth and income are observed not just at the point of production (fishing), but also along the fisheries value chain. Depending on the type of fish that is caught, fish can be consumed within the

Fig. 3.1 Nets of fishers in Darocotan Bay, Philippines. (Photo credit: Katherine Jack)

household, bartered or given away, or sold at local urban, larger urban or international markets. A characteristic feature of many of these fish sales is that of the personalised economic relationship, which in essence involves a regular or favoured trading relationship (commonly referred to as *suki*). In theory, this offers mutual benefits for fishers and buyers: buyers are assured of regular supply, especially useful when supply is low, whereas fishers are assured of a regular buyer for their products (Ruddle, 2011). While local *suki* relationships are common, the intensity of the relationship can be increased when longer commodity chains and higher levels of capital are involved. In cases where fish are transported internationally (e.g., squid, live reef fish), buyers with greater resources finance fishing trips, the construction of boats with engines, and in some cases even the personal expenses (e.g., school fees, food) of fishers. Fishers are bound to sell their catch to the buyer who has financed them, and those who are financed receive lower prices than those who do not rely on a financier and are able to independently sell their fish.

The relationships between fishers and buyers in *suki* relationships are highly variable, but marked by power relations (Russell, 1987). This is particularly so when there are large differences in wealth between the two parties, and when a significant amount of credit is provided. In the case of export fisheries such as live fish and squid, for example, the local buyers who can afford to extend credit to fishers are either agents of exporters based in Manila, or local entrepreneurs with significant assets, and in many cases are politically well connected. In these cases, the relationship between the fisher and the trader closely resembles that of the patron and the client common in analyses of the Philippine political system, where the patron supplies the client with protection and security, and the client provides a loyal vote. Many traders higher up the value chain, such as the agents of exporters based in the provinces or the exporters themselves, have connections with ethnic Chinese, for example, through marriage.

Important in these contexts is the ability of fishers to actively work social relations to obtain relevant introductions to buyers and/or financiers. The personalised nature of the relationship is apparent in the common use of fictive kinship terms (e.g., *kuya*, older brother; *tatay*, father), and references to specific cultural values such as *pakinabangan* (reciprocity), *apa* (empathy) and *hiya* (shame) (Turgo, 2016). While patron–client relations in fisheries are sometimes criticised because some fishers become bound up in long-term debts (e.g., Padilla et al. 2003), fishers actively seek to turn these hierarchical relationships to their advantage. When negotiating the terms of such relationships, fishers will often appeal to strongly held moral values about the obligations of the well-off to take pity on the poor (*awa*) and of the poor's 'right to survive' (Szanton, 1972). Fishers' claims in these ongoing relationships (e.g., financing the purchase of a boat, or gaining a position on a fishing vessel) are, therefore, situated within a broader cultural context (Fabinyi, 2012). Thus, fishers are economically differentiated in relation to the production process itself, and along the value chain, in ways that intersect with other forms of differentiation such as migrant status, and with broader cultural values.

The diversity of the roles associated with fisheries production and trade and the ways in which these roles link in with overlapping class and status relationships mean that a 'fishing livelihood' can only be understood in relation to its position within the local grid of social relationships. Therefore, as with all livelihoods, relations of power are a crucial—and frequently overlooked—component of livelihoods (De Haan & Zoomers, 2005; Scoones, 2009). Often, these power relations are located beyond the geographically defined community in which a fisher lives: as Pauwelussen (2015: 332) notes in relation to Indonesian maritime peoples, 'affinity and loyalty follow translocal relations of kinship, credit, and debt rather than the borders of a village or island'.

These forms of differentiation and power relations are reflected not only in everyday operations of fishing livelihoods, but also become particularly visible through governance interventions. Where MPAs are located may have greater consequences for some groups of fishers and not others, and the decision-making processes by which these locations are finalised can reflect these power dynamics. For example, inshore fishers without motorised boats—and who tend to be poorer and with less political clout—have been disadvantaged in the site selection of MPAs (Eder, 2005; Fabinyi, 2012).

In the Philippines, the importance of power relations also comes into play in discussions of various types of illegal fishing. Illegal fishing is a contested term that can encompass a diverse set of fishing activities, from the use of 'active' gears in municipal waters such as beach seines, to the incursion of large-scale commercial fishing vessels in municipal waters, to the use of air compressors and of destructive gears such as cyanide and dynamite. Elsewhere in the coastal Philippines, Knudsen (2012, 2013) has shown that more recent migrants with lower status were subtly excluded from the benefits of marine conservation projects, and were more frequently blamed for illegal fishing incidents. In other parts of the coastal Philippines, blast fishers (Galvez et al., 1989) and commercial fishers illegally fishing within coastal waters (Segi, 2014a) have effectively integrated into the local community and are tolerated because of their power deriving from the significant economic benefits they distribute. In other cases, well-resourced illegal fishers are alleged to simply bribe government officials to allow them to operate (Fabinyi, 2012). In all these cases, formal governance of fishing livelihoods interacts with and is subsequently shaped by relationships of power and wealth on the ground.

GENDER IN FISHING LIVELIHOODS IN OCEANIA

All over the world the participation of women in fishing livelihoods is overlooked (e.g., see Ram, 1991). It is difficult to study women's roles in fisheries because data on fisheries are rarely disaggregated by sex of fisher (Harper et al., 2020). A systematic review of the literature on gender in fisheries found that in many parts of the world, including Africa and the Pacific Islands, women tend to fish close to shore, often gleaning for invertebrates, while men tend to fish further away from shore and catch more vertebrates (Kleiber et al., 2015). The paper found that people, including women, tend to see women's fishing as assisting men's fishing or as assisting with household incomes or food production, rather than as being important activities in their own right. For example, women in Kiribati looking after young children may take them swimming in shallow water, but also take a net and catch fish for dinner at the same time. When fishing is part-time or is for subsistence rather than for cash, it is often omitted from fisheries and census data collection. Moreover, gleaning as a method and invertebrates as catch are also often omitted from fisheries data collection (Gopal et al., 2020; Weeratunge et al., 2010). These factors combined render women's fishing invisible. For example, in the Pacific Island country of Wallis and Futuna a fisheries official told a visiting fisheries consultant that women do not fish in Wallis and Futuna. Later the consultant and the official had a lunch of shellfish together. When asked who collected the shellfish, the official answered that women did, but that this was not 'fishing' (Barclay et al., 2019: 3).

Fisheries researchers have found that in the countries of Oceania the proportion of women fishing for food and livelihoods ranges from around 20 per cent in some countries to around 50 per cent in others (Harper et al., 2013). One study in the Marovo Lagoon in Solomon Islands found that over 80 per cent of women fish or glean for invertebrates, and 84 per cent of women who fish target finfish rather than invertebrates (Rabbitt et al., 2019). Another study found that in Fiji women fishers play critical roles in food security and livelihoods (Thomas et al., 2020). However, the prevailing assumption that women do not fish is so pervasive that many women who fish do not see themselves as fishers, and data collection systems continue to omit the kinds of part-time, near-shore, invertebrate-focused fishing women most often do. For example, since around 2015 the Household Income and Expenditure Survey conducted in Pacific Island countries has asked women and men whether they fish for food and incomes. When the first-year data were collected only 8 per cent of women reported that they worked as fishers (Secretariat of the Pacific Community [SPC], 2017). Recognising that this figure was improbably low, the following year the Report Card noted 'women's participation in fisheries is underestimated by [the survey] … with one country estimate being that no women are employed in fisheries (range 0% to 17%)' (2018: 3). The figures were improved the year after that, with women being reported as making up 18 per cent of the fisheries workforce (range 8–38 per cent) (2019).

Gender blindness causes problems for sustainable resource management. For example, women's fishing often includes gleaning in the intertidal zone, which is rarely included in fisheries monitoring, meaning that monitoring is failing to pick up important information about human effects on marine ecosystems (Kleiber et al., 2015). Fijian women mud crab fishers have taken matters into their own hands and established a community-based fisheries management plan to address overfishing (Giffin et al., 2019). Vanuatu fisheries managers (pers. comm. with Barclay, May 2017) related that recent efforts to be more inclusive in their community consultations have revealed new and important factors for coastal fisheries management. In the past they had not been aware of practices used in octopus fishing, because it is mainly women who fish for octopus in shallow waters. Fisheries managers had been used to talking only to men about fishing and focused on the types of fishing men did, usually further out from shore using boats. In the 2010s they started talking with women about their fishing. They discovered that the main way women fished for octopus was through using metal bars to break or overturn the corals octopuses hide in, which is a destructive fishing practice. If they had only talked to men, as in the past, they would have remained unaware that destructive fishing practices were being used to fish for octopus, and, thus, not addressed the problem.

In a community in Solomon Islands, research around fishing practices found that a local MPA was less effective than it could have been due to gender blindness in creating the MPA. Women were not effectively consulted and the MPA was placed over fishing grounds women commonly used. Thus, the MPA establishment process lacked legitimacy in women's eyes, and obeying the rules of the MPA would have made their lives more difficult due to having to go further afield to fish, so some women were flouting the rules of the MPA by fishing in the no-take zone (Rohe et al., 2018).

Another example of overlooking the resource sustainability implications of near-shore fishing for invertebrates as part of a livelihood activity mainly conducted by women is that of shell money in the Langalanga Lagoon in Solomon Islands (see Fig. 3.2). Various kinds of customary exchange valuables have long been part of cultural life in Melanesia. The shell money produced in the Langalanga Lagoon continues to be valued

Fig. 3.2 Shell money production in Langlalanga Lagoon, Solomon Islands. (Photo credit: Kate Barclay)

in modern life, with some types of pieces being used for weddings and other ceremonial purposes, and simpler pieces used as casual jewellery. White beads are made from a shell called in local language *kakadu* (*Anadara granosa*) and black beads used at the end of shell money strings are from *kurila* (*Atrina vexillum*). Some beads are heated to bring out their colour, including red beads from *romu* (*Chama pacifica*) and orange beads *ke'e* (*Beguina semiorbiculata*). With continuing demand for shell money the shells used have been depleted in the Langalanga area, so are supplied from further afield in Solomon Islands. By 2014 traders were reporting that overfishing might also be occurring outside Langalanga, because the size of shells was getting smaller (Barclay et al., 2018). Solomon Islands has some important export fisheries for invertebrates, such as sea cucumber, trochus and pearl shells, so has long been monitoring and attempting to tackle overfishing through periodic export bans. However, the shells used for shell money are not on the list of species for monitoring. We do not know why this is the case, but in line with the global tendency to overlook women in fisheries, we speculate that it could be because shell money production in Langalanga is largely women's activity.

Another problem that arises from the invisibility of women's fishing is that training or funding for fisheries-related livelihood activities is usually targeted at men. For example, a project to support community-based aquaculture in Pacific Island countries included a specific gender focus and engaged gender specialists in some of their activities (Jimmy et al., 2019). They found in Fiji and Samoa that women and men tended to identify men as the fish farmers, and not women, despite the fact that women were doing much of the day-to-day work operating fish farms. A handful of farms in Fiji run by women were exceptions to this rule. In the past the tendency to assume men were fish farmers and women were not meant that men received any training and were the ones involved in discussions with the fisheries agency supporting aquaculture, where decisions were made. The project concluded that fisheries agencies should have capacity building for gender analysis and gender mainstreaming so that they would be better able to observe where women were involved and include them appropriately in interventions (Jimmy et al., 2019). These findings are similar to case studies of aquaculture in Bangladesh, where women's involvement is also underestimated, showing that with careful project design women may be empowered (Choudhury et al., 2017).

Even when fisheries agencies and other organisations involved in funding or implementing fisheries livelihood projects recognise that women are important players and should be involved, ingrained gender roles, norms and expectations can make it difficult to meaningfully engage women. It is not as simple as just inviting women to the meetings where projects will be planned and decisions made—although that is a necessary first small step. After the WorldFish office in Solomon Islands had been through a process of capacity building to tackle gender transformation in their work, they started to implement this in their work with fishing communities. In one of their early attempts they found difficulties even in inviting women to meetings when they went out to villages. Their established way of inviting people to meetings was by letter or through other communication with village leaders, who were men and who were not accustomed to passing on news about such upcoming meetings to women in their communities. Thus, the fieldworkers had to organise meetings with women once they arrived, by walking around the village and seeing which women could participate at short notice. For their part women were not accustomed to participating in such meetings, so eliciting their participation was hard, and focus groups took much longer than planned. Women in Solomon Islands have less formal schooling than men, so it was more difficult to translate some concepts that were not part of their daily lives, such as 'nutrition' (Jones et al., 2014). Pursuing their aim for livelihoods activities to be gender transformative, over the years since that first attempt WorldFish Solomon Islands staff have accumulated learning about how to effectively engage village women in livelihood activities. These include many practical points such as ensuring that women are not spending a great deal of time cooking for visitor meetings, and that meetings work around any caring responsibilities women may have (Gomese et al., 2020; Lawless et al., 2017).

Finally, gender intersects with other forms of social marginalisation. Some communities, including the women in them, see that the most pressing social issues affecting their livelihoods are related to factors other than gender, and they want those other issues dealt with first in any interventions. For example, in a study of the contributions of tuna fisheries to coastal communities, one of the study areas was a fishing village specialising in handlining for yellowfin tuna near Gizo in Solomon Islands. The members of this village are ethnic I-Kiribati, having migrated from the former British territory of the Gilbert and Ellis Islands during the twentieth century, but they are Solomon Islands citizens (McClean et al., 2019).

The broader study found that across the study locations in Solomon Islands and Indonesia the main factors affecting the distribution of contributions from fisheries to livelihoods were gender, ethnicity (especially migrant status) and socio-economic status. However, when the Gizo participants (women and men) were asked about gender as a factor, they responded that they were not concerned about the gender relations in fishing. They wanted the project to report on their marginalisation as a migrant community, and for any interventions to address those problems (McClean et al., 2019).

Conclusion

While the idea of a 'fisher' is a convenient shorthand term, it conceals both the diversity of practices associated with a fishing livelihood, and the other forms of identity that interact with a fishing livelihood. Different types of fishers and others whose livelihoods are based on the fish value chain have different sets of interests, and frequently these roles correspond with other forms of identity such as gender or ethnicity. Such forms of difference are also organised hierarchically in relations of power (i.e., particular groups of fishers and social groups tend to be marginalised, while others are not). While in some contexts such forms of marginalisation are arguably becoming increasingly well recognised (e.g., in relation to gender), in other contexts marginalisation can be more difficult to unpack and recognise (e.g., relationships between recent migrants and long-term residents).

The implication for fisheries governance is that these forms of social differentiation influence how people respond to or are affected by any new fisheries governance initiative. For example, a new MPA or the imposition of a closed season is mediated and influenced by these social institutions. Failure to be inclusive can simply mean that the governance intervention fails to achieve its objectives, as in the case of the MPA in Solomon Islands (Rohe et al., 2018), while inclusive resource management can enable fisheries agencies to better manage resources, as in the case of octopus fishing in Vanuatu. In some cases, failure to be inclusive can lead to significant social effects and generate social tensions (e.g., Segi, 2014b). While the challenges for organisations and policymakers in recognising and addressing social difference are significant, and require long-term sustained effort, the potential for genuinely improving fishing livelihoods is correspondingly substantial. For example, WorldFish as an organisation has spent many years embedding a gender transformative approach to its work. It can now point to measurable outcomes from interventions, whereby women have increased choice regarding income activities in seafood supply chain and control over the income they generate, with corresponding positive effects for their families (Cole et al., 2020).

REFERENCES

Barclay, K., McClean, N., Foale, S., Sulu, R., & Lawless, S. (2018). Lagoon livelihoods: Gender and shell money in Langalanga, Solomon Islands. *Maritime Studies, 17*, 199–211. https://doi.org/10.1007/s40152-018-0111-y

Barclay, K., Leduc, B., Mangubhai, S., Vunisea, A., Namakin, B., Teimarane, M., & Leweniqila, L. (2019). Module 1: Introduction. In K. Barclay, B. Leduc, S. Mangubhai, & C. Donato-Hunt (Eds.), *Pacific handbook for gender equity and social inclusion in coastal fisheries and aquaculture*. Pacific Community (SPC).

Béné, C. (2003). When fishery rhymes with poverty: A first step beyond the old paradigm on poverty in small-scale fisheries. *World Development, 31*(6), 949–975. https://doi.org/10.1016/S0305-750X(03)00045-7

Béné, C., & Friend, R. M. (2011). Poverty in small-scale fisheries: Old issue, new analysis. *Progress in Development Studies, 11*(2), 119–144. https://doi.org/10.1177/146499341001100203

Bernstein, H. (2010). *Class dynamics of agrarian change* (Vol. 1). Kumarian Press.

Choudhury, A., McDougall, C., Rajaratnam, S., & Park, C. M. Y. (2017). *Women's empowerment in aquaculture: Two case studies from Bangladesh*. FAO; WorldFish.

Cole, S. M., Kaminski, A. M., McDougall, C., Kefi, A. S., Marinda, P. A., Maliko, M., & Mtonga, J. (2020). Gender accommodative versus transformative approaches: A comparative assessment within a post-harvest fish loss reduction intervention. *Gender, Technology and Development, 24*(1), 48–65. https://doi.org/10.1080/09718524.2020.1729480

Coulthard, S. (2008). Adapting to environmental change in artisanal fisheries—Insights from a South Indian Lagoon. *Global Environmental Change, 18*(3), 479–489. https://doi.org/10.1016/j.gloenvcha.2008.04.003

De Haan, L., & Zoomers, A. (2005). Exploring the frontier of livelihoods research. *Development and Change, 36*(1), 27–47. https://doi.org/10.1111/j.0012-155X.2005.00401.x

Dressler, W. H., & Fabinyi, M. (2011). Farmer gone fish'n? Swidden decline and the rise of grouper fishing on Palawan Island, the Philippines. *Journal of Agrarian Change, 11*(4), 536–555. https://doi.org/10.1111/j.1471-0366.2011.00309.x

Dressler, W. H., & Turner, S. (2008). The persistence of social differentiation in the Philippine uplands. *The Journal of Development Studies, 44*(10), 1450–1473. https://doi.org/10.1080/00220380802360966

Eder, J. F. (2005). Coastal resource management and social differences in Philippine fishing communities. *Human Ecology, 33*(2), 147–169. https://doi. org/10.1007/s10745-005-2430-Z

Eder, J. F. (2008). *Migrants to the coasts: Livelihood, resource management, and global change in the Philippines.* Nelson Education.

Fabinyi, M. (2012). *Fishing for fairness.* ANU Press.

Fabinyi, M., Foale, S., & Macintyre, M. (2015). Managing inequality or managing stocks? An ethnographic perspective on the governance of small-scale fisheries. *Fish and Fisheries, 16*(3), 471–485. https://doi.org/10.1111/faf.12069

Fabinyi, M., Dressler, W. H., & Pido, M. D. (2017). Fish, trade and food security: Moving beyond 'availability' discourse in marine conservation. *Human Ecology, 45*(2), 177–188. https://doi.org/10.1007/s10745-016-9874-1

Galvez, R. E., Hingco, T. G., Bautista, C., & Tungpalan, M. T. (1989). Sociocultural dynamics of blast fishing and sodium cyanide fishing in two fishing villages in the Lingayen Gulf area. In G. Silvestre, E. Miclat, & C. Thia-Eng (Eds.), *Towards sustainable development of the coastal resources of Lingayen Gulf, Philippines* (pp. 43–62). ICLARM Conference Proceedings 17. International Center for Living Aquatic Resources Management.

Geertz, C. (1973). *The interpretation of cultures.* Basic Books.

Giffin, A. L., Naleba, M., Fox, M., & Mangubhai, S. (2019). Women fishers in Fiji launch a mud crab management plan for their fishery. *SPC Women in Fisheries Information Bulletin, 30*, 20–23.

Gomese, C., Siota, F., Ride, A., & Kleiber, D. (2020). Reflections on integrating gender-sensitive facilitation techniques in fieldtrip reports. *SPC Women in Fisheries Information Bulletin, 32*, 28–30.

Gopal, N., Hapke, H. M., Kusakabe, K., Rajaratnam, S., & Williams, M. J. (2020). Expanding the horizons for women in fisheries and aquaculture. *Gender, Technology and Development, 24*(1), 1–9. https://doi.org/10.1080/0971852 4.2020.1736353

Hall, D., Hirsch, P., & Li, T. M. (2011). *Powers of exclusion: Land dilemmas in Southeast Asia.* University of Hawai'i Press.

Harper, S., Zeller, D., Hauzer, M., Pauly, D., & Sumaila, U. R. (2013). Women and fisheries: Contribution to food security and local economies. *Marine Policy, 39*(1), 56–63. https://doi.org/10.1016/j.marpol.2012.10.018

Harper, S., Adshade, M., Lam, V. W. Y., Pauly, D., & Sumaila, U. R. (2020). Valuing invisible catches: Estimating the global contribution by women to small-scale marine capture fisheries production. *PLoS One, 15*(3), 1–17. https://doi.org/10.1371/journal.pone.0228912

Hornborg, A., Clark, B., & Hermele, K. (Eds.). (2013). *Ecology and power: Struggles over land and material resources in the past, present and future* (Vol. 18). Routledge.

Jimmy, R., Pickering, T., & Tioti, B. (2019). *Improving community-based aquaculture in Fiji, Kiribati, Samoa and Vanuatu.* Final Report, FIS/2012/076. ACIAR.

Jones, C., Schwarz, A. M., Sulu, R., & Tikai, P. (2014). *Foods and diets of communities involved in inland aquaculture in Malaita Province, Solomon Islands.* Program Report, AAS-2014-30. CGIAR Research Program on Aquatic Agricultural Systems.

Kerkvliet, B. J. T. (1990). *Everyday politics in the Philippines: Class and status relations in a Central Luzon village.* University of California Press.

Kleiber, D., Harris, L. M., & Vincent, A. C. J. (2015). Gender and small-scale fisheries: A case for counting women and beyond. *Fish and Fisheries, 16*(4), 547–562. https://doi.org/10.1111/faf.12075

Knudsen, M. (2012). Fishing families and cosmopolitans in conflict over land on a Philippine island. *Journal of Southeast Asian Studies, 43*(3), 478–499. https://doi.org/10.1017/S0022463412000355

Knudsen, M. (2013). Beyond clientelism: Neighbourhood leaders on a Philippine Island. *Anthropological Forum, 23*(3), 242–265. https://doi.org/10.1080/00664677.2013.812032

Knudsen, M. (2016). Poverty and beyond: Small-scale fishing in overexploited marine environments. *Human Ecology, 44*(3), 341–352. https://doi.org/10.1007/s10745-016-9824-y

Lawless, S., Doyle, K., Cohen, P., Eriksson, H., Schwarz, A. -M., Teioli, H., McDougall, C., et al. (2017). *Considering gender: Practical guidance for rural development initiatives in Solomon Islands.* Program Brief 2017–22. WorldFish.

Li, T. M. (2000). Articulating Indigenous identity in Indonesia: Resource politics and the tribal slot. *Comparative Studies in Society and History, 42*(1), 149–179. https://doi.org/10.1017/S0010417500002632

Lowe, C. (2002). Who is to blame? Logics of responsibility in the live reef food fish trade in Sulawesi, Indonesia. *SPC Live Reef Fish Information Bulletin, 10,* 7–16.

McCall Howard, P. (2012). Sharing or appropriation? Share systems, class and commodity relations in Scottish fisheries. *Journal of Agrarian Change, 12*(2–3), 316–343. https://doi.org/10.1111/j.1471-0366.2011.00355.x

McClean, N., Barclay, K., Fabinyi, M., Adhuri, D. S., Sulu, R. J., & Indrabudi, T. (2019). *Assessing tuna fisheries governance for community wellbeing: Case studies from Indonesia and Solomon Islands.* University of Technology Sydney. Retrieved February 5, 2021, from https://www.uts.edu.au/about/faculty-arts-and-social-sciences/research/fass-research-projects/assessing-governance-tuna

Mills, D., Béné, C., Ovie, S., Tafida, A., Sinaba, F., Kodio, A., Lemoalle, J., et al. (2011). Vulnerability in African small-scale fishing communities. *Journal of International Development, 23*(2), 308–313. https://doi.org/10.1002/jid.1638

Padilla, J. E., Mamauag, S., Braganza, G., Brucal, N., Yu, D., & Morales, A. (2003). *Sustainability assessment of the live reef-fish for food industry in Palawan, Philippines.* Philippines: World Wildlife Fund.

Pauwelussen, A. P. (2015). The moves of a Bajau middlewoman: Understanding the disparity between trade networks and marine conservation. *Anthropological Forum, 25*(4), 329–349. https://doi.org/10.1080/00664677.2015.1054343

Peet, R., & Watts, M. (Eds.). (1996). *Liberation ecologies: Environment, development, social movements*. Routledge.

Rabbitt, S., Lilley, I., Albert, S., & Tibbetts, I. R. (2019). What's the catch in who fishes? Fisherwomen's contributions to fisheries and food security in Marovo Lagoon, Solomon Islands. *Marine Policy, 108*, 103667. https://doi.org/10.1016/j.marpol.2019.103667

Ram, K. (1991). *Mukkuvar women: Gender, hegemony, and capitalist transformation in a south Indian fishing community*. Zed Books.

Ribot, J. C., & Peluso, N. L. (2003). A theory of access. *Rural Sociology, 68*(2), 153–181. https://doi.org/10.1111/j.1549-0831.2003.tb00133.x

Rocheleau, D., Thomas-Slayter, B., & Wangari, E. (Eds.). (1996). *Feminist political ecology: Global issues and local experience*. Routledge.

Rohe, J., Schlüter, A., & Ferse, S. C. A. (2018). A gender lens on women's harvesting activities and interactions with local marine governance in a South Pacific fishing community. *Maritime Studies, 17*, 155–162. https://doi.org/10.1007/s40152-018-0106-8

Ruddle, K. (2011). 'Informal' credit systems in fishing communities: Issues and examples from Vietnam. *Human Organization, 70*(3), 224–232. https://doi.org/10.17730/humo.70.3.v4810k37717h9g01

Russell, S. D. (1987). Middlemen and moneylending: Relations of exchange in a highland Philippine economy. *Journal of Anthropological Research, 43*(2), 139–161.

Russell, S., & Poopetch, M. (1990). Petty commodity fishermen in the inner gulf of Thailand. *Human Organization, 49*(2), 174–187. https://doi.org/10.17730/humo.49.2.u26301427q364380

Sahlins, M. (1978). Culture as protein and profit. *New York Review of Books, 25*(18), 45–53.

Scoones, I. (2009). Livelihoods perspectives and rural development. *The Journal of Peasant Studies, 36*(1), 171–196. https://doi.org/10.1080/03066150902820503

Secretariat of the Pacific Community. (2017). *Coastal fishery report card 2017*. Pacific Community (SPC). Retrieved February 5, 2021, from http://www.spc.int/DigitalLibrary/Doc/FAME/Brochures/SPC_2016_Coastal_Fishery_Report_Card.pdf

Secretariat of the Pacific Community. (2018). *Coastal fishery report card 2018*. Pacific Community (SPC). Retrieved February 5, 2021, from http://www.spc.int/DigitalLibrary/Doc/FAME/Brochures/SPC_2018_Coastal_Fishery_Report_Card.pdf

Secretariat of the Pacific Community. (2019). *Coastal fishery report card 2019*. Pacific Community (SPC). Retrieved February 5, 2021, from http://www.spc.int/DigitalLibrary/Doc/FAME/Brochures/SPC_2019_Coastal_Fishery_Report_Card.pdf

Segi, S. (2014a). 'Losing at sea, winning on land': A case study of Philippine small-scale and industrial fisher resource competition. *Society & Natural Resources, 27*(12), 1227–1241. https://doi.org/10.1080/08941920.2014.948237

Segi, S. (2014b). Protecting or pilfering? Neoliberal conservationist marine protected areas in the experience of coastal Granada, the Philippines. *Human Ecology, 42*(4), 565–575. https://doi.org/10.1007/s10745-014-9669-1

Sidel, J. T. (1999). *Capital, coercion, and crime: Bossism in the Philippines.* Stanford University Press.

Stacey, N., Steenbergen, D. J., Clifton, J., & Acciaioli, G. L. (2018). Understanding social wellbeing and values of small-scale fisheries amongst the Sama-Bajau of archipelagic Southeast Asia. In D. S. Johnson, T. G. Acott, N. Stacey, & J. Urquhart (Eds.), *Social wellbeing and the values of small-scale fisheries* (pp. 97–123). Springer.

Szanton, M. C. B. (1972). *A right to survive: Subsistence marketing in a lowland Philippine town.* Pennsylvania State University Press.

Thomas, A. S., Mangubhai, S., Fox, M., Lalavanua, W., Meo, S., Miller, K., Naisilisili, W., et al. (2020). Valuing the critical roles and contributions of women fishers to food security and livelihoods in Fiji. *Women in Fisheries Information Bulletin, 31,* 22–29.

Turgo, N. (2016). 'Laway lang ang kapital' (Saliva as capital): Social embeddedness of market practices in brokerage houses in the Philippines. *Journal of Rural Studies, 43,* 83–93. https://doi.org/10.1016/j.jrurstud.2015.11.001

Weeratunge, N., Snyder, K. A., & Sze, C. P. (2010). Gleaner, fisher, trader, processor: Understanding gendered employment in fisheries and aquaculture. *Fish and Fisheries, 11*(4), 405–420. https://doi.org/10.1111/j.1467-2979.2010.00368.x

CHAPTER 4

Fisheries Governance

Abstract This chapter examines the role that governance plays in shaping fishing livelihoods. This includes formal government regulation as well as other factors that shape fishing, such as markets, buyer requirements and social norms. Institutional arrangements serve as a key component of fishing livelihoods, by prescribing the conditions under which fishing livelihoods operate. In this chapter we sketch out some of the trends in fisheries governance across parts of the Asia-Pacific, before discussing examples in Australia and Indonesia.

Keywords Fisheries governance • Neoliberalism • Resource nationalism • Indonesia • Australia

This chapter turns to an examination of the role that governance plays in shaping fishing livelihoods. While fishing has long been perceived as a classic example of the 'tragedy of the commons'[1] the reality is that most if not all fishing livelihoods are significantly affected by some sort of institutional arrangements (Kooiman et al., 2005; Ostrom, 1990). These institutional arrangements serve as a key component of fishing livelihoods, by prescribing the conditions under which fishing livelihoods operate. This chapter

---

[1] Or more accurately, the tragedy of open access (Bromley & Cernea, 1989; Ostrom, 1990).

M. Fabinyi, K. Barclay, *Asia-Pacific Fishing Livelihoods*, https://doi.org/10.1007/978-3-030-79591-7_4

sketches out some of the trends in fisheries governance across parts of the Asia-Pacific, before moving on to discuss examples in Australia and Indonesia.

Adapting a definition from Hall et al. (2011: 16) in relation to land, here we consider governance as the formal and informal rules that govern access and exclusion over fisheries resources. While state-based governance is most commonly thought of when governance is discussed, many other wider sets of social institutions regulate access to and exclusion from fisheries resources (Bromley, 1992; Kooiman et al., 2005; Ostrom, 1990). In addition to government, civil society and private sector actors are increasingly involved in governance partnerships. Other institutions include informal social norms, conventions or negotiated arrangements, such as those between different groups of fishers surrounding gear use, or between fishers and traders surrounding financing and credit.

States of the Asia-Pacific, as elsewhere, have profoundly different levels of governance capacity and resources devoted to fisheries governance. In many small-scale fisheries across the Asia-Pacific, fisheries are effectively 'self-governed'. In other words, the formal reach of the state has limited purchase, and access to fisheries resources is governed through customary and/or informal institutions that overlap with many of the social institutions described in Chap. 3. For example, in much of the Pacific forms of customary marine tenure can regulate access to marine space along clan lines (Carrier, 1981), or restrict access to waters for a certain period of time (Cohen & Foale, 2013; Hviding, 1996). Other variations of customary institutions have been well documented for South-East Asia (Ruddle & Satria, 2010) and South Asia (Coulthard, 2011). An important point to note in this context is that these customary institutions were largely not designed to manage marine resources biologically or ecologically, but to regulate social access (i.e., restricting fishing access to neighbouring groups) (Foale et al., 2011). In contemporary times, many of these customary institutions have been significantly transformed or now coexist with more formal state regulations in conditions of legal pluralism (Bavinck, 2018; Bavinck et al., 2013; Lau et al., 2020).

Historically, the most fundamental model of fisheries governance by states in the Asia-Pacific has been one of resource nationalism (Koch & Perreault, 2019), where states have explicitly aimed to expand fisheries production and trading. As Campling and Havice (2018: 88) point out in their insightful historical analysis of national seafood production systems, 'national seafood systems promoted volume [of extraction] to ensure the

reproduction of domestic capital, sustain new industrial societies by pro-
viding cheap food for workers and their families, and extend[ed] geopo-
litical influence'. This process was reflected throughout the vast drive
towards industrial expansion in much of Asia in the postwar period, discussed in Chap. 2 (see Butcher, 2004; Christensen, 2014). In the Pacific,
a related variety of resource nationalism took place through a domestica-
tion model, mostly for tuna. This involved leveraging good natural
resources and tariff advantages to compensate for distance from trade
routes, lack of infrastructure and high labour costs (Barclay &
Cartwright, 2008).

In many developing countries of the Asia-Pacific there has been a shift
towards the concept of 'co-management', centred around the principle of
shared responsibility for management between the state and resource
users, as well as the participation of other stakeholders such as civil society
groups (Evans et al., 2011; Ratner et al., 2012). The underlying objective
was to improve both the effectiveness of fisheries resource management
and the legitimacy of the state through the active participation of resource
users. In practice, co-management models vary on a continuum from centralised, where government undertakes most functions, through consulta-
tive and collaborative, to delegated models, where fishers undertake most
governance functions (Pomeroy & Berkes, 1997).

The outcomes of co-management initiatives in the Asia-Pacific are
highly variable (Quimby & Levine, 2018; Sunderlin & Gorospe, 1997).
In Indonesia and the Philippines, for example, co-management is strongly
influenced by other social institutions, including kinship, ethnicity, or customary leadership structures, which affect whose interests are prioritised
and who effectively participates in decision-making (Eder, 2005;
Steenbergen, 2016). Across South-East Asia, the development of co-
management has in many cases been supported by foreign donors, linked
in with the rise of community-based management programmes or accom-
panied by the decentralisation of fisheries management to local governments (Christie et al., 2005; Courtney & White, 2000). For example, in
the Philippines decentralisation led to the demarcation of designated spatial zones for small-scale and industrial fishers (with varying degrees of
enforcement). Australia has a centralised model, whereby representatives
of industry are involved in advisory groups that meet with fisheries agency
staff and review documents but have no decision-making power. Many
fishers feel profoundly disempowered in this system (Barclay et al., 2020;
King & O'Meara, 2019). Japan has had a delegated model with some

fisheries management decisions made through fisheries cooperatives, which during the twentieth century acted well to protect fisher interests, but was not successful at preventing overfishing in key fisheries such as tuna (Barclay & Koh, 2008).

Common to both richer and poorer countries has been a shift towards EBFM, which aims to replace conventional fisheries science based only on the target species, with understanding of a fishery's effects on the broader ecosystem (Pikitch et al., 2004). For example, some trawl fisheries catch a great deal of other species in addition to the target species, and may damage habitat through dragging trawling gear along the bottom of the sea. A conventional fisheries science approach would look only at the stocks of target species. An EBFM approach would look at all the species being affected by the trawling, and the effect of dragging on the ocean floor. Despite widespread acceptance by governments and scientists internationally since the early 1990s that fisheries should be managed as part of ecosystems (Pikitch et al., 2004), EBFM has largely not been implemented. EBFM constitutes a radical change from existing single species-based management, and it has been unclear how the shift to EBFM may feasibly be achieved (Barclay, 2016). Closely linked with the emergence of marine spatial planning, EBFM is also often associated with the implementation of MPAs. The long-time horizons and variable nature of fishery benefits generated by fish spillover from MPAs, combined with the short-term effects on fishing grounds, mean that the implementation of MPAs has had mixed results for fisher livelihoods (Ban et al., 2019; Gill et al., 2019; Segi, 2014).

More recently has been a shift to what has been broadly termed as 'private' or 'market-based governance', based on the idea of market actors taking a leading role in governing for environmental sustainability (Bush & Oosterveer, 2019; Groeneveld et al., 2017). This idea has been most notably applied to fisheries in the case of eco-certification and labelling (e.g., the MSC and the sustainable seafood movement). Under this model of governance, transparency is implemented through traceability documentation (Bailey et al., 2016). While the direct involvement of the private sector as leading actors in fisheries governance is a relatively new phenomenon, it builds on the logic of economic rationalism, or neoliberalism, that has driven much fisheries governance for several decades, especially in richer countries.

Neoliberalism as a particular type of governance, especially of economic activity, has become increasingly widespread since the 1980s. Neoliberalism is an all-embracing kind of term, used in some cases to refer to specific policy mixes involving deregulation, privatisation and use of market mechanisms in the policy sphere, but also to refer more diffusely to a form of governmentality (McCormack, 2017a). In marine governance neoliberalism has taken shape from a particular vision dating from the 1950s of oceans as commons with inherent problems of overfishing and overcapacity (Mansfield, 2004). In this logic it is human nature to overexploit commons resources, to competitively race to fish and innovate technologically in that race to fish, causing both overfishing and overcapacity (McCormack, 2017a). Converting ocean commons to private property and then using market mechanisms to allocate access to the resource is seen as a way to harness the profit motive to achieve conservation objectives and improve economic efficiency (Mansfield, 2004; McCormack, 2017a). Public access to resources has been limited by turning commons into private property that can be controlled and traded (Mansfield, 2007). In fisheries management neoliberal privatisation and market mechanisms have been brought together in the form of individual transferable quotas (ITQs) (Mansfield, 2004; McCormack, 2017a, 2017b; Pinkerton, 2017).

ITQs build on existing foundations in fisheries science of stock assessments generating a TAC—an amount of the target species that may be harvested annually. In ITQs the TAC is divided into quota shares, which are often allocated among fishers according to catch history or levels of investment. Each quota holder is allowed to catch up to the limit of its quota (usually a tonnage). This is called an 'output' control on fisheries, in that the control is on what comes out of the fishery, as opposed to 'input' controls on what goes into a fishery, such as controls on gear (e.g., net mesh size) or temporal limits to the fishing season. Fishing quotas may be traded. Less efficient operators tend to sell their quota to more efficient operators in a process that reduces the overall number of quota owners and increases the profitability of the quota owners who remain. However, this does not necessarily mean fishers themselves experience economic improvements. For example, the British Columbia halibut fishery is often hailed as an ITQ economic success, but 79 per cent of the quota was leased—most fishers did not own quota themselves. For quota-leasing fishers the economics of operational and leasing costs in relation to fish prices was much less favourable (Pinkerton & Edwards, 2009).

ITQ proponents argue that having property rights in fisheries encourages quota owners to see their wealth as dependent on the health of fish stocks, therefore, encouraging stewardship of the resource (McCormack,

2017a). There is no empirical evidence that ITQs are correlated with or cause norm change regarding environmental stewardship (Hoshino et al., 2019). Indeed, ITQ systems have been found to encourage high grading and dumping, poaching and under-reporting (McCormack, 2017b; Pinkerton, 2017).

The neoliberal ITQ model has not universally been seen as a good template for fisheries management. For example, in Japan the preferred model has been for co-management with fisheries cooperatives, whereby the cooperatives were largely responsible for controlling the fishing activities of members (Barclay & Epstein, 2013; Barclay & Koh, 2008). The Japanese co-management with cooperatives model has not been particularly successful at curbing overfishing, with two infamous examples being whaling and bluefin tuna fisheries (Epstein & Barclay, 2013), but it has arguably been effective in terms of managing conflicts between fishers (Matsuda, 1987) and preserving fishing livelihoods in rural parts of Japan (Barclay & Koh, 2008). Nevertheless, neoliberal fisheries management ideas have eventually started to infiltrate Japan's fisheries management and into that of many other countries that have hitherto resisted ITQs (McCormack, 2017a: 32). Yet, at the same time, challenges to neoliberalism in fisheries policy are also gaining ground around the world (Pinkerton, 2017).

ITQs may only be feasible in wealthy countries, because they require a high level of state involvement in ascertaining TACs and administering quota systems. Moreover, ITQs embody a 'wealth' rather than a 'welfare' orientation regarding fisheries (Béné et al., 2010). ITQs explicitly aim to accumulate the wealth of a fishery among fewer participants (Hoshino et al., 2019; Pinkerton, 2017). Even early studies noted that ITQs tend to concentrate ownership, causing smaller operators to leave the industry (Connor & Alden, 2001; McCay, 1995). Studies have found there is often decreased employment in fisheries where ITQs are implemented, and that high quota prices act as a barrier preventing fishers from becoming quota owners (Hoshino et al., 2019).

In this introductory section of the chapter we have briefly highlighted several of the most prominent models for fisheries governance that are widely adopted around the world, such as customary institutions, resource nationalism, co-management, EBFM, market-based fisheries governance and neoliberalism. An important point to note about all governance models is that they are not neutral technical interventions, but represent particular ideas about the world, based on valuations of people and the environment (Li, 2007). As formal regulations become implemented in practice, they interact with other social institutions such as culture and social norms that, in turn, have been generated over time. In this chapter we consider two cases of different governance models operating in very different contexts and trace their implications for fishing livelihoods.

FISHERIES GOVERNANCE IN AUSTRALIA

Fisheries governance in Australia until the end of the 1980s was largely open access, aimed at generating jobs and increasing food supplies. There was overfishing and had been since the earliest days of colonisation (Wilkinson, 1997). For example, settlers in South Australia harvested oysters so excessively they wiped out many oyster beds in the 1800s (Wallace-Carter, 1987). In line with a general move towards ecologically sustainable development around 1990, fisheries management moved towards biologically sustainable fishing, preventing overfishing. In terms of economic goals for fisheries, the Australian Government (2019) aims to 'maximise the net economic return to the Australian community' (Commonwealth of Australia, 2017: 2) from the use of fisheries resources, and to have 'cost recovery'—meaning have fishers pay for fisheries management (Patterson et al., 2020). Australian fisheries governance since the 1990s has favoured profitable fisheries over other economic considerations such as job creation, and implemented ITQs fairly widely.

One of the prominent models of fishing in Australia has been that of small-scale fishing using a single boat with diversified methods across species, gear, locations and markets, to respond to fluctuations in environmental and market conditions (Minnegal & Dwyer, 2008; Voyer et al., 2016) (see Fig. 4.1). It was a viable business model to manage the inherent risks in near-shore and estuarine fisheries that have great natural variation in stock availability (Barclay et al., 2020; Voyer et al., 2016). By the late 1990s fisheries management in Australia had moved in a direction that discouraged this kind of fishing in favour of larger-scale specialist operations (Minnegal & Dwyer, 2008). This was part of a wave of similar policies pursued internationally to discourage smaller diversified operators in favour of larger-scale specialist operators (Hilborn et al., 2001). Diversified small operators are generally not as efficient as larger operations with equipment specialised for a specific fishery, and Australian fisheries management has worked to reduce fleet sizes by pushing out inefficient operators (Connor & Alden, 2001).

Fig. 4.1 Fisher with freshly caught eel in Victoria, Australia. (Photo credit: Impress Photography)

Small-scale diversified fishers were also discouraged because of the ecologically sustainable development approach wanting to reduce 'latent capacity' or 'latent effort'. Diversified operators keep rights in fisheries they do not often use so that when the environmental and market conditions suit they can work in that fishery, but most of the time they do not. In practical terms there is not overfishing because the rights are not fully used, but there is the potential that if fishers did fully use all of their rights at once there could be overfishing. As fisheries managers want to reduce this risk, they have moved to get rid of little-used rights (Barclay et al., 2020; Minnegal & Dwyer, 2008).

Regulation in favour of economic efficiency and eliminating latent effort has, thus, discouraged diversified fishing. Alongside the regulation has been a discourse that delegitimises fishing operations that are low profit. The concepts of net economic returns to the community and cost recovery mean that fisheries 'should' be able to pay a resource rent back to government and cover fisheries management costs. If they are not profitable enough to do this then it is argued that they should not have access to the resource. Fishers who are not very profitable have been stigmatised as 'lifestyle' fishers rather than as 'business-oriented' fishers. Simply supporting a family or employing oneself in one's chosen vocation is no longer seen as a legitimate use of commons fisheries resources (Minnegal & Dwyer, 2008; Voyer et al., 2016).

Australian fisheries management since the 1990s has been neoliberal, working towards privatising resource access rights in ITQs and allocating them through market mechanisms (Bichler et al., 2019; Organisation for Economic Co-operation and Development, n.d.) as the best way to meet the objectives of ecologically sustainable development and economic efficiency in Commonwealth and state fisheries legislation (Commonwealth of Australia, 2017; Minnegal & Dwyer, 2008). Not all fisheries were put under ITQ management in the 1990s or subsequently, but even non-ITQ fisheries were strongly influenced by the governance mode prioritising the prevention of overfishing over the social benefits of fisheries, and favouring profitable operating styles over ones that generate more jobs.

Some large-scale, high-value fisheries have flourished with this type of management. For example, the highly migratory southern bluefin tuna were thoroughly overfished internationally by the 1980s. Prior to the fishery coming under quota management, 136 fishing boats were in the southern bluefin tuna fishery across three Australian states. ITQs were introduced into the southern bluefin fishery in 1984 and well over half of the boats left the fishery within two years (Campbell et al., 2000). By the mid-1990s all southern bluefin quota had been consolidated into 13 companies based in Port Lincoln in South Australia. There the industry developed a ranching system that enabled the fishery to recover economically, becoming so successful that the regional economy around Port Lincoln was boosted by the industry. Port Lincoln became famous for having a high per-capita number of millionaire 'tuna barons', but the industry also generated a great deal of jobs. In the financial year 2016–2017 tuna farming in Port Lincoln and flow-on activities such as processing and transport amounted to 856 full-time equivalent jobs, and each of these jobs is estimated to have created an additional 2.13 jobs elsewhere in the state economy (Econsearch, 2018).

However, in other Australian ITQ fisheries fishers' livelihoods have not fared so well. For example, in Tasmania's abalone fishery much of the quota is owned by investors who lease the quota to fishers for harvesting. Profits in the fishery go to the quota owners rather than the fishers, and the fishers have no security of access to the resource. Quota prices are prohibitively high, which combined with quota rarely coming onto the market means a barrier to new entrants (Hoshino et al., 2019).

The ITQ model is now being rolled out in lower-value, small-scale, multi-species, multi-gear fisheries that have made up the majority of fishing in parts of Australia, and many in the industry are concerned about negative effects on livelihoods. ITQs were implemented broadly in New South Wales from 2017. The effects on income and employment are not clear because no data are collected on these aspects of fisheries in most of Australia, but a majority of fishers say the reforms have damaged their livelihoods, and their wellbeing appears to have suffered from the reform process (Barclay et al., 2020). Fishers in Queensland and South Australia facing imminent ITQ reforms similarly fear that many operators will be forced out of the industry (McClean, Voyer, et al., 2019b; Sutton Sutton, 2020).

A neoliberal approach to fisheries, including ITQs, is also at odds with the values, rights, knowledge systems and social practices of Indigenous peoples in Australia (Lalancette, 2017; Schnierer & Egan, 2015). ITQs have been embedded into Māori fishing rights by the New Zealand Government, which has led to widening income inequality between general tribe members and those with decision-making rights over quota, and reduced Māori people's rights to fish as a livelihood (McCormack, 2013).

Over the same time period as the Australian Government took a neoliberal approach to fisheries management and shifted from supporting livelihoods to preventing overfishing, conservationist perspectives have also permeated the Australian public consciousness more broadly. Science communication and conservation organisation media campaigns spread awareness of the problems of overfishing internationally. Images of overfishing were received by the Australian public in ways that resulted in a cultural devaluing of fisheries as a livelihood. Fishing, previously somewhat stigmatised because of its working-class status, has become even more stigmatised, with fishers often treated by the public as environmental 'rapers and pillagers' (Kearney, 2013).

Australian fishers are sometimes spat on and have rocks thrown at them, their vehicles and fishing equipment are vandalised, and strangers yell at them, claiming that they are destroying the marine environment (King, 2018; Voyer et al., 2016). Some people fish at night to avoid encounters with the public at the wharf, and build high fences to hide fishing gear in their yards. Some children of fishers lie at school about their parents' jobs (Voyer et al., 2016). There is a budding food localism movement, but in general fishers are not valued in Australian society as food producers. Over 70 per cent of seafood consumed in Australia is imported (Kearney, 2013).

In Australia the public has been primed with the narrative 'they take all the fish' by the recreational fishing movement. Recreational fishing has long been a popular pastime in Australia, and complaints by recreational fishers that their sport is ruined by professional fishers taking too many fish were published in Sydney newspapers as early as the 1860s (Clark, 2017). Since the 1970s recreational fishing has grown in popularity and since the 2000s recreational fishing lobby groups have leveraged their large constituency to wield a great deal of political influence (Voyer et al., 2017).

Lack of public support for professional fishing, environmental movements focused on establishing reserves where fishing is banned, and recreational fishing lobbying activities have caused professional fishers to lose access to fishing grounds. In 2001 professional fishers had access to 113 water bodies in New South Wales, of which 24 supplied 95 per cent of all fish caught professionally in the state. By 2012 professional fishing was banned or restricted in 15 of those 24, due to zoning of those water bodies as 'recreational fishing havens' or MPAs (Stevens et al., 2012: 5). In Victoria the recreational fishing lobby has succeeded in having the state government ban professional fishing in places such as Port Philip Bay, despite the scientific evidence indicating that those fisheries were biologically sustainable, and with no evidence that removing professional fishing would improve recreational fishing outcomes (King & O'Meara, 2019).

What will happen to fishing livelihoods in Australian in the future? Not many young people are willing or able to enter the fishing industry for a range of reasons outlined above—the high cost of buying quota, lost fishing grounds, the stigma fishing has acquired as environmentally damaging—and other reasons to do with difficult regulations and high production costs (Abernethy et al., 2020; Barclay et al., 2020; King et al., 2019; Minnegal & Dwyer, 2008; Shaw et al., 2011; Voyer et al., 2016). Fishing livelihoods remain but are increasingly in corporate operations rather than the small-scale diverse fisheries that characterised much Australian fishing in the past. There is also a group of family businesses capitalising on growing food localism among consumers and moving up the value chain into direct sales, for example, through farmers markets (Abernethy et al., 2020; Voyer et al., 2016). With market disruptions due to COVID-19 there has been reorientation of export value chains to domestic markets, and interest in local production of food has strengthened. Therefore, there is hope for fishers who weather the COVID-19 storm and the longer-term pressures of neoliberal governance.

FISHERIES GOVERNANCE IN INDONESIA

Fisheries governance in Indonesia is very different to that of Australia. Indonesia is a much larger producing state, the second largest fishery producer by volume in the world after China, whereas Australia is around fiftieth. Indonesia's fisheries are diverse, mostly informal and small scale, spread across thousands of beaches, ports and inland waterways. Fishing is foundational to the food supply and livelihoods of coastal communities in Indonesia's very large population. Around 12 per cent are below the poverty line and 27 per cent are characterised as being vulnerable to slipping into poverty. Poverty rates are disproportionately high in fishing communities (World Bank, 2015). As a middle-income country Indonesia has less government resources available for fisheries management than high-income countries like Australia.

For centuries fishing has been a mainstay livelihood activity along the vast stretches of coast and inland waterways of the islands that became the modern state of Indonesia. This included food for local consumption as well as the trade in dried marine products, such as *trepang* (dried sea cucumbers) that have long been traded around South-East Asia and to China. As in the rest of South-East Asia, European colonial arrangements and then Japanese fishing companies were influential in the establishment of industrial fishing in the twentieth century.

In the 1970s Japan retreated somewhat from international fishing, due to rising wages and other production costs in the Japanese fleet, including the advent of fishing access payments due to the establishment of 200 nm EEZs under UNCLOS (Barclay, 2014). This left space for Indonesian industrial fishing companies to develop. In 1975 the Indonesian Government created the tuna longline company Perikanan Samodra Besar (known also as PSB or PERSERO), which early on worked with the Japanese fleet, but then moved on to operate independently. In the 1980s, the company expanded, and the Indonesian Government financially supported other industrial fishing companies (Morgan & Staples, 2006).

State governance of fisheries in Indonesia has largely been oriented to industrial fishing and processing, taking a resource nationalist approach of increasing production to increase economic benefits. Indonesia's support for domestic industrial fisheries was part of a tide of resource nationalism among former colonies in the 1970s. Newly decolonised states hoped to use control over their own resources to establish a New International Economic Order, wherein former colonies were not subordinated to former coloniser states. They took inspiration from what OPEC countries managed to achieve with their oil resources. In fisheries this coincided with the negotiations that led to UNCLOS, with states taking economic control of waters to 200 nm from their coast (i.e., EEZs), meaning distant water fishing states like Japan had to negotiate access to fishing grounds they had previously used for free. Therefore, in fisheries, the 1970s trend for resource nationalism took the form of developing domestic industrial fisheries, requiring distant water fishing states to invest in joint ventures and onshore processing, or to pay fees in return for access to EEZs (Schurman, 1998).

Indonesia's 2004 *Fisheries Act* has nine objectives for fisheries management, one of which is about guaranteeing the sustainability of fish resources, and the other eight are about using the resources for economic goals, including improved living standards for small-scale fishers, government revenue, population nutrition, and supporting industrial fishing and processing. However, to date, Indonesian state management of fisheries has mainly facilitated industrial expansion in fishing and processing, rather than monitoring fish stocks or applying limits to fisheries for the purpose of sustaining fish stocks (California Environmental Associates [CEA], 2018; Sunoko & Huang, 2014). The expansion of fishing has meant more jobs on fleets, in processing and in fish trading. It is difficult to say exactly how many people have incomes from fishing in Indonesia, but it is likely in the millions. For example, in the area around Bitung, which is a major industrial tuna fishing and processing hub in Eastern Indonesia, it is estimated there are at least 12,000 small-scale tuna fishers (Sukarsih et al., 2019) (see Fig. 4.2). There would be thousands more on the industrial fleet and in the processing plants.

Indonesian government capacity to monitor and regulate fisheries is limited as a developing country with a huge population and enormous fisheries spread over thousands of islands, as well as a complex governance system between national, provincial and municipal level agencies (Cabral et al., 2018). Overall fisheries policy and licensing for vessels over 30 GT, and management of maritime areas outside 12 nm, sits with national offices of the Ministry of Marine Affairs and Fisheries. Areas within 12 nm are the responsibility of provincial government offices of the Ministry.

Fig. 4.2   Freshly caught yellowfin tuna. (Photo credit: Katherine Jack)

Small vessels must be registered but do not pay license fees. Indonesia has had limited control over fisheries and poor fisheries data, in part because of the huge diverse fleet operating from thousands of mostly informal landing sites (Cabral et al., 2018; CEA 2018). Given that Indonesia is such a significant fishing country, since the 1990s other countries sharing migratory stocks such as tuna have worked with Indonesia through regional fisheries management organisation ('RFMO') processes to try to improve data (Hanich et al., 2010).

Indonesian fisheries governance shifted somewhat with the appointment of Minister of Fisheries Susi Pudjiastuti (2014–2019). With a dual agenda to prevent illegal, unreported and unregulated (IUU) fishing and to replace foreign investment, Minister Susi presided over the implementation of policies that saw Filipino fishers and vessels leave Indonesian waters, including exploding confiscated fishing vessels (Cabral et al., 2018; McClean, Barclay, et al., 2019a). In addition to removing foreign vessels, the anti-IUU measures also decreased the amount of tuna from Indonesian waters being sent overseas for processing, with the aim to increase the supply of fish for Indonesian processors—although in the short term the removal of Filipino vessels caused fish supplies for the industrial processors to drop (McClean, Barclay, et al., 2019a). Prior to Susi's tenure, the government largely neglected small-scale fisheries, and here again Susi broke new ground. She announced that the Filipino large-scale tuna fishing vessels banned from 2014 would be replaced with 3325 new small- to medium-scale vessels to be built in Indonesia with government funding and given to Indonesian fishers (McClean, Barclay, et al., 2019a).

Minister Susi's tenure was intensely controversial in the Indonesian seafood industry, but she had popular support and remained in the post for five years. She was eventually replaced in 2019 by Edhy Prabowo, who was less obvious in supporting any particular governance direction. In November 2020 Minister Prabowo was arrested for alleged corruption regarding exports of lobster seed (Dao, 2020).

The Indonesian Government has not so far implemented neoliberal fisheries management tools such as ITQs. It does not have TACs, which are foundational for ITQs. The Ministry of Marine Affairs and Fisheries continues to work towards strengthening data collection and stock assessment, aiming to develop harvest strategies for key fisheries such as tuna (Hoshino et al., 2020). In line with fisheries management internationally, Indonesia is also moving towards ecosystem-based management, with co-management between government and industry (Muawanah et al., 2018).

There has been some adoption of certification for export markets, with effects that constitute a limited form of market-based governance. One issue with market-based certification is that it requires extensive documentation and verification at each stage of the supply chain. Such 'audit culture' was developed in high-income countries and is highly unsuited to small-scale informal Indonesian fisheries, where documentation is not a normal part of life (Bush et al., 2013). Therefore, small-scale fishers entering into certification require intermediaries to assist with the documentation; these are usually the trading companies that buy the fish for export.

On the biological front, several fisheries are in MSC assessment or are moving towards it as part of fishery improvement programmes, which are mostly supported by philanthropic organisations (Levine et al., 2020). Some small-scale handline tuna fisheries exporting to US markets entered fair-trade certification in the mid-2010s. Fair-trade fishers are paid a price premium as a community development fund, which can be used for mutually beneficial things such as fishing safety equipment or community projects. For the purposes of traceability catches are documented, which creates an opportunity to monitor catches (previously small-scale catches were not recorded). Over 700 fishers around Ambon now provide catch data that feed into provincial-level fisheries data collection used in decision-making Fisheries Co-Management Committees (McClean, Barclay, et al., 2019a). In recent years the use of forced labour (slavery) in Indonesian fisheries and seafood processing has become a prominent issue but as yet there is no market-based measure that effectively tackles labour conditions.

Diverse forms of customary practices and laws relating to marine spaces at the local level remain important institutions across the country. For example, among Sama-Bajau fishers there are prohibitions against harming or hunting whale sharks (Stacey et al., 2012). In Aceh Province there are local 'sea commanders' (*panglima laot*) who are responsible for enforcing laws relating to the sea, such as dispute resolution among fishers, access to fishing grounds, and management of mooring sites and ports (Nurasa et al., 1993; Wilson & Linkie, 2012). In other parts of Indonesia, such as in Maluku and Sulawesi, customary laws such as spatial and temporal closures (e.g., for trochus) of marine resources (*sasi*) have been well documented (Satria & Adhuri, 2010; Thorburn, 2000). Typically, such customary practices and laws form part of wider social institutions, and have varying levels of codification, recognition by government, and enforcement. They can be highly dynamic and, increasingly, environmental NGOs are assessing and harnessing these customary practices and laws for their potential to contribute to marine conservation and sustainable resource management (McLeod et al., 2009; Zerner, 1994).

CONCLUSION

Different modes of fisheries governance have evolved across the Asia-Pacific at multiple scales of governance. Highly diverse, they represent competing logics of how to manage people and resources. Customary forms of governance have long regulated access to marine resources, embedded within the wider social worlds of culture and social relationships. As the power and capacity of many states grew after World War II, resource nationalism became the default policy for many states in the Asia-Pacific, as they sought to extract the maximum benefits from their oceans through extensive support for commercial fishing industries. More recently, concerns about overfishing and ecological degradation have also become key factors underlying fisheries governance in many countries, as have forms of governance that seek to harness, and maximise, the financial aspects of fisheries. Many high-income countries have adopted neoliberal fisheries policy approaches, which focus on aggregate generation of wealth, rather than focusing on welfare, considering the distribution of livelihoods across groups in society. In contrast, movements that emphasise the roles, knowledge, expertise and rights of local communities and fishers have manifested in forms of governance such as co-management and human rights-based approaches (Allison et al., 2012; Jentoft et al., 2017).

There is an implicit assumption in some of the fisheries literature that policymaking is a technical process, that it is possible to attain the right outcomes if only the right governance institutions are put in place. Yet, policies are always implemented in particular contexts—they interact with pre-existing forms of governance, and with the social and political-economic contexts discussed in Chaps. 2 and 3. Understanding how the policymaking *process* itself evolves is just as significant for assessing outcomes as the particular contents of any policy. For example, in the Australian case, the intersection of recreational fishing, conservation discourse and neoliberal fisheries management converged to govern fishing livelihoods in particular ways.

Ultimately, governance institutions that unfold in particular contexts shape the kinds of livelihoods that are on offer and to whom. In the Indonesian case, state regulation has meant that Filipino fishers were pushed out by new regulations, while informal and customary institutions also enable some groups to fish and not others. Certification initiatives require collaborators to handle the paperwork, which leads to fishers being dependent on those partners, and fishers who do not have such partnerships are excluded from certification (McClean, Barclay, et al., 2019a). In the Australian case, the neoliberal approach to governing fisheries has led away from small-scale diversified fishing livelihoods towards more corporate, specialised operations.

In the context of the blue economy—where interest by state, market and civil society actors in the governance and use of the oceans is rapidly expanding (Voyer et al., 2018)—the influence of different ideas about governance on fishing livelihoods will only increase. Fishing livelihoods have been under pressure to demonstrate their environmental sustainability and the transparency and traceability of their operations, and to comply with intensified environmental governance such as MPAs. As the generation of economic wealth and the protection of the natural environment emerge as powerful themes in the blue economy, ensuring that fishing livelihoods are adequately represented and included in governance is a crucial challenge for policymakers.

REFERENCES

Abernethy, K., Barclay, K., McIlgorm, A., Gilmour, P., McClean, N., & Davey, J. (2020). *Victoria's fisheries and aquaculture: Economic and social contributions.* Research project. FRDC 2017–092. University of Technology Sydney.

Allison, E. H., Ratner, B. D., Åsgård, B., Willmann, R., Pomeroy, R., & Kurien, J. (2012). Rights-based fisheries governance: From fishing rights to human rights. *Fish and Fisheries, 13*(1), 14–29. https://doi.org/10.1111/j.1467-2979.2011.00405.x

Australian Government. (2019). *Cost recovery implementation statement 2018–19.* Australian Fisheries Management Authority. Retrieved February 5, 2021, from https://www.afma.gov.au/sites/default/files/uploads/2018/06/AFMA-Cost-Recovery-Implementation-Statement-CRIS-2018-19_.pdf

Bailey, M., Bush, S. R., Miller, A., & Kochen, M. (2016). The role of traceability in transforming seafood governance in the global south. *Current Opinion in Environmental Sustainability, 18,* 25–32. https://doi.org/10.1016/j.cosust.2015.06.004

Ban, N. C., Gurney, G. G., Marshall, N. A., Whitney, C. K., Mills, M., Gelcich, S., Tran, T. C., et al. (2019). Well-being outcomes of marine protected areas. *Nature Sustainability, 2*(6), 524–532. https://doi.org/10.1038/s41893-019-0306-2

Barclay, K. (2014). History of industrial tuna fishing in the Pacific Islands. In J. Christensen & M. Tull (Eds.), *Historical perspectives of fisheries exploitation in the Indo-Pacific* (pp. 153–171). MARE Series, Vol. 12. Springer.

Barclay, K. (2016). Futures of governance: Ecological challenges and policy myths in tuna fisheries. In J. P. Marshall & L. H. Connor (Eds.), *Environmental change and the world's futures: Ecologies, ontologies and mythologies* (pp. 65–80). Routledge.

Barclay, K., & Cartwright, I. (2008). *Capturing the wealth from tuna: Case studies from the Pacific.* ANU Press.

Barclay, K., & Epstein, C. (2013). Securing fish for the nation: Food security and governmentality in Japan. *Asian Studies Review, 37*(2), 215–233. https://doi.org/10.1080/10357823.2013.769498

Barclay, K., & Koh, S.-H. (2008). Neoliberal reforms in Japan's tuna fisheries? A history of government-business relations in a food-producing sector. *Japan Forum, 20*(2), 139–170. https://doi.org/10.1080/09555800802047475

Barclay, K., Davila, F., Kim, Y., McClean, N., & Mcilgorm, A. (2020). *Economic analysis and social and economic monitoring following the NSW commercial fisheries business adjustment program.* Institute for Sustainable Futures, University of Technology Sydney. Retrieved February 5, 2021, from https://www.dpi.nsw.gov.au/__data/assets/pdf_file/0007/1256128/Economic-analysis-and-Social-and-Economic-monitoring-following-the-NSW-Commercial-Fisheries-Business-Adjustment-Program.pdf

Bavinck, M. (2018). Legal pluralism, governance, and the dynamics of seafood supply chains—Explorations from South Asia. *Maritime Studies, 17*(3), 275–284. https://doi.org/10.1007/s40152-018-0118-4

Bavinck, M., Johnson, D., Amarasinghe, O., Rubinoff, J., Southwold-Llewellyn, S., & Thomson, K. T. (2013). From indifference to mutual support—A comparative analysis of legal pluralism in the governing of south Asian fisheries. *The European Journal of Development Research, 25*(4), 621–640. https://doi.org/10.1057/ejdr.2012.52

Béné, C., Hersoug, B., & Allison, E. H. (2010). Not by rent alone: Analysing the pro-poor functions of small-scale fisheries in developing countries. *Development Policy Review, 28*(3), 325–358. https://doi.org/10.1111/j.1467-7679.2010.00486.x

Bichler, M., Fux, V., & Goeree, J. K. (2019). Designing combinatorial exchanges for the reallocation of resource rights. *Proceedings of the National Academy of Sciences of the United States of America, 116*(3), 786–791. https://doi.org/10.1073/pnas.1802123116

Bromley, D. W. (1992). The commons, common property, and environmental policy. *Environmental and Resource Economics, 2*(1), 1–17. https://doi.org/10.1007/BF00324686

Bromley, D. W., & Cernea, M. M. (1989). *The management of common property natural resources: Some conceptual and operational fallacies.* Discussion Paper no. 57. World Bank.

Bush, S. R., & Oosterveer, P. (2019). *Governing sustainable seafood.* Routledge.

Bush, S. R., Toonen, H., Oosterveer, P., & Mol, A. P. J. (2013). The 'devils triangle' of MSC certification: Balancing credibility, accessibility and continuous improvement. *Marine Policy, 37*(1), 288–293. https://doi.org/10.1016/j.marpol.2012.05.011

Butcher, J. G. (2004). *The closing of the frontier: A history of the marine fisheries of Southeast Asia, c. 1850–2000.* Institute of Southeast Asian Studies.

Cabral, R. B., Mayorga, J., Clemence, M., Lynham, J., Koeshendrajana, S., Muawanah, U., Costello, C., et al. (2018). Rapid and lasting gains from solving illegal fishing. *Nature Ecology & Evolution, 2*(4), 650–658. https://doi.org/10.1038/s41559-018-0499-1

California Environmental Associates. (2018). *Trends in marine resources and fisheries management in Indonesia: A review.* CEA. Retrieved September 30, 2020, from https://www.packard.org/wp-content/uploads/2018/08/Indonesia-Marine-Full-Report-08.07.2018.pdf

Campbell, D., Brown, D., & Battaglene, T. (2000). Individual transferable catch quotas: Australian experience in the Southern Bluefin tuna fishery. *Marine Policy, 24*(2), 109–117. https://doi.org/10.1016/S0308-597X(99)00017-2

Campling, L., & Havice, E. (2018). The global environmental politics and political economy of seafood systems. *Global Environmental Politics, 18*(2), 72–92. https://doi.org/10.1162/glep_a_00453

Carrier, J. G. (1981). Ownership of productive resources on Ponam Island, Manus province. *Journal. Société des Océanistes, 37*(72), 205–217. https://doi.org/10.3406/jso.1981.3061

Christensen, J. (2014). Unsettled seas: Towards a history of marine animal populations in the central Indo-Pacific. In J. Christensen & M. Tull (Eds.), *Historical perspectives of fisheries exploitation in the Indo-Pacific* (pp. 13–39). MARE Series, Vol. 12. Springer.

Christie, P., Lowry, K., White, A. T., Oracion, E. G., Sievanen, L., Pomeroy, R., Pollnac, R., et al. (2005). Key findings from a multidisciplinary examination of integrated coastal management process sustainability. *Ocean & Coastal Management, 48*(3–6), 468–483. https://doi.org/10.1016/j.ocecoaman.2005.04.006

Clark, A. (2017). *The catch: The story of fishing in Australia* (1st ed.). National Library of Australia.

Cohen, P. J., & Foale, S. J. (2013). Sustaining small-scale fisheries with periodically harvested marine reserves. *Marine Policy, 37*, 278–287. https://doi.org/10.1016/j.marpol.2012.05.010

Commonwealth of Australia. (2017). *Commonwealth fisheries policy statement.* Department of Agriculture and Water Resources. Retrieved February 5, 2021, from https://www.agriculture.gov.au/sites/default/files/sitecollectiondocuments/fisheries/domestic/cwlth-fisheries-policy-statement.pdf

Connor, R., & Alden, D. (2001). Indicators of the effectiveness of quota markets: The south east trawl fishery of Australia. *Marine and Freshwater Research, 52*(4), 387–397. https://doi.org/10.1071/MF99164

Coulthard, S. (2011). More than just access to fish: The pros and cons of fisher participation in a customary marine tenure (*Padu*) system under pressure. *Marine Policy, 35*(3), 405–412. https://doi.org/10.1016/j.marpol.2010.11.006

Courtney, C. A., & White, A. T. (2000). Integrated coastal management in the Philippines: Testing new paradigms. *Coastal Management, 28*(1), 39–53. https://doi.org/10.1080/089207500263639

Dao, T. (2020, December 2). *Indonesian fisheries minister arrested in baby lobster export probe.* SeafoodSource. Retrieved February 5, 2021, from https://www.seafoodsource.com/news/supply-trade/indonesian-fisheries-minister-arrested-in-baby-lobster-export-probe

Econsearch. (2018). *The economic contribution of aquaculture in the south Australian state and regional economies, 2016/17.* Adelaide, South Australia. Retrieved December 3, 2020, from https://pir.sa.gov.au/aquaculture/publications/_nocache

Eder, J. F. (2005). Coastal resource management and social differences in Philippine fishing communities. *Human Ecology, 33*(2), 147–169. https://doi.org/10.1007/s10745-005-2430-Z

Epstein, C., & Barclay, K. (2013). Shaming to 'green': Australia–Japan relations and whales and tuna compared. *International Relations of the Asia-Pacific, 13*(1), 95–123. https://doi.org/10.1093/irap/lcs019

Evans, L., Cherrett, N., & Pemsl, D. (2011). Assessing the impact of fisheries co-management interventions in developing countries: A meta-analysis. *Journal of Environmental Management, 92*(8), 1938–1949. https://doi.org/10.1016/j.jenvman.2011.03.010

Foale, S., Cohen, P., Januchowski-Hartley, S., Wenger, A., & Macintyre, M. (2011). Tenure and taboos: Origins and implications for fisheries in the Pacific. *Fish and Fisheries, 12*(4), 357–369.

Gill, D. A., Cheng, S. H., Glew, L., Aigner, E., Bennett, N. J., & Mascia, M. B. (2019). Social synergies, tradeoffs, and equity in marine conservation impacts. *Annual Review of Environment and Resources, 44*, 347–372. https://doi.org/10.1146/annurev-environ-110718-032344

Groeneveld, R. A., Bush, S. R., & Bailey, M. (2017). Private governance of ocean resources. In P. A. L. D. Nunes, L. E. Svensson, & A. Markandya (Eds.), *Handbook on the economics and management of sustainable oceans* (pp. 416–428). Edward Elgar Publishing.

Hall, D., Hirsch, P., & Li, T. M. (2011). *Powers of exclusion: Land dilemmas in Southeast Asia*. University of Hawai'i Press.

Hanich, Q., Tsamenyi, M., & Parris, H. (2010). Sovereignty and cooperation in regional Pacific tuna fisheries management: Politics, economics, conservation and the vessel day scheme. *Australian Journal of Maritime and Ocean Affairs, 2*(1), 2–15. https://doi.org/10.1080/18366503.2010.10815650

Hilborn, R., Maguire, J.-J., Parma, A. M., & Rosenberg, A. A. (2001). The precautionary approach and risk management: Can they increase the probability of successes in fishery management? *Canadian Journal of Fisheries and Aquatic Sciences, 58*(1), 99–107. https://doi.org/10.1139/f00-225

Hoshino, E., van Putten, I., Pascoe, S., & Vieira, S. (2019). Individual transferable quotas in achieving multiple objectives of fisheries management. *Marine Policy, 113*, 103744. https://doi.org/10.1016/j.marpol.2019.103744

Hoshino, E., Hillary, R., Davies, C., Satria, F., Sadiyah, L., Ernawati, T., & Proctor, C. (2020). Development of pilot empirical harvest strategies for tropical tuna in Indonesian archipelagic waters: Case studies of skipjack and yellowfin tuna. *Fisheries Research, 227*, 105539. https://doi.org/10.1016/j.fishres.2020.105539

Hviding, E. (1996). *Guardians of Marovo lagoon: Practice, place, and politics in maritime Melanesia* (Vol. 14). University of Hawai'i Press.

Jentoft, S., Chuenpagdee, R., Barragán-Paladines, M. J., & Franz, N. (Eds). (2017). *The small-scale fisheries guidelines: Global implementation*. MARE Series, Vol. 14. Springer.

Kearney, R. (2013). Australia's out-dated concern over fishing threatens wise marine conservation and ecologically sustainable seafood supply. *Open Journal of Marine Science, 3*(2), 55–61. https://doi.org/10.4236/ojms.2013.32006

King, T. J. (2018, October 19). *Project regard.* Youtube video. Retrieved February 5, 2021, from https://www.youtube.com/watch?v=e-QQqx3qGck

King, T. J., & O'Meara, D. (2019). 'The people have spoken': How cultural narratives politically trumped the best available science (BAS) in managing the port Phillip Bay fishery in Australia. *Maritime Studies, 18*(1), 17–29. https://doi.org/10.1007/s40152-018-0097-5

King, T. J., Abernethy, K., Brumby, S., Hatherell, T., Kilpatrick, S., Munksgaard, K., & Turner, R. (2019). *Sustainable fishing families: Developing industry human capital through health, wellbeing, safety and resilience.* Research Project. FRDC 2016-400. Canberra: Fisheries Research and Development Corporation, Deakin University, Western District Health Service, University of Tasmania, and University of Exeter.

Koch, N., & Perreault, T. (2019). Resource nationalism. *Progress in Human Geography, 43*(4), 611–631. https://doi.org/10.1177/0309132518781497

Kooiman, J., Jentoft, S., Bavinck, M., & Pullin, R. (Eds.). (2005). *Fish for life: Interactive governance for fisheries.* Amsterdam University Press.

Lalancette, A. (2017). Creeping in? Neoliberalism, indigenous realities and tropical rock lobster (*kaiar*) management in Torres Strait, Australia. *Marine Policy, 80*, 47–59. https://doi.org/10.1016/j.marpol.2016.02.020

Lau, J. D., Cinner, J. E., Fabinyi, M., Gurney, G. G., & Hicks, C. C. (2020). Access to marine ecosystem services: Examining entanglement and legitimacy in customary institutions. *World Development, 126*, 104730. https://doi.org/10.1016/j.worlddev.2019.104730

Levine, M., Thomas, J. B., Sanders, S., Berger, M. F., Gagern, A., & Michelin, M. (2020). *2020 global landscape review of fishery improvement projects.* CEA Consulting. Retrieved February 5, 2021, from https://oursharedseas.com/oss_downloads/2020-global-landscape-review-of-fishery-improvement-projects/

Li, T. M. (2007). *The will to improve: Governmentality, development, and the practice of politics.* Duke University Press.

Mansfield, B. (2004). Neoliberalism in the oceans: 'Rationalization', property rights, and the commons question. *Geoforum, 35*(3), 313–326. https://doi.org/10.1016/j.geoforum.2003.05.002

Mansfield, B. (2007). Privatization: Property and the remaking of nature–society relations introduction to the special issue. *Antipode, 39*(3), 393–405. https://doi.org/10.1111/j.1467-8330.2007.00532.x

Matsuda, Y. (1987). Postwar development and expansion of Japan's tuna fishery. In D. Doulman (Ed.), *Tuna issues and perspectives in the Pacific Islands region* (pp. 71–91). East-West Center.

McCay, B. J. (1995). Social and ecological implications of ITQs: An overview. *Ocean & Coastal Management, 28,* 3–22.

McClean, N., Barclay, K., Fabinyi, M., Adhuri, D. S., Sulu, R. J., & Indrabudi, T. (2019a). *Assessing tuna fisheries governance for community wellbeing: Case studies from Indonesia and Solomon Islands.* University of Technology Sydney. Retrieved February 5, 2021, from https://www.uts.edu.au/about/faculty-arts-and-social-sciences/research/fass-research-projects/assessing-governance-tuna

McClean, N., Voyer, M., Davila, F., Barclay, K., Cunningham, R., & Schnierer, S. (2019b). *Analytical report on historical factors and barriers, thematic analysis, typology of stakeholders, and social network analysis.* Report Prepared for the Queensland Government, Department of Agriculture and Fisheries. Retrieved January 1, 2021 from https://www.publications.qld.gov.au/dataset/queensland-sustainable-fisheries-strategy/resource/1fb9427f-ecae-4113-a9c3-66ee58fa2696

McCormack, F. (2013). Commodities and gifts in New Zealand and Hawaiian fisheries. In F. McCormack & K. Barclay (Eds.), *Engaging with capitalism: Case studies from Oceania* (pp. 53–81). Emerald Group Publishing Limited. https://doi.org/10.1108/S0190-1281(2013)0000033005

McCormack, F. (2017a). *Private oceans: The enclosure and marketisation of the seas.* Pluto Press and University of Chicago Press.

McCormack, F. (2017b). Sustainability in New Zealand's quota management system: A convenient story. *Marine Policy, 80,* 35–46. https://doi.org/10.1016/j.marpol.2016.06.022

McLeod, E., Szuster, B., & Salm, R. (2009). *Sasi* and marine conservation in Raja Ampat, Indonesia. *Coastal Management, 37*(6), 656–676. https://doi.org/10.1080/08920750903244143

Minnegal, M., & Dwyer, P. D. (2008). Managing risk, resisting management: Stability and diversity in a southern Australian fishing fleet. *Human Organization, 67*(1), 97–108. https://doi.org/10.17730/humo.67.1.x38g60k463p26855

Morgan, G. R., & Staples, D. J. (2006). Tuna longlining, poling and purse seining. In G. R. Morgan & D. J. Staples (Eds.), *The history of industrial marine fisheries in Southeast Asia.* UN FAO Regional Office for Asia and the Pacific. Retrieved January 1, 2021, from http://www.fao.org/3/AG122E00.htm#Contents

Muawanah, U., Yusuf, G., Adrianto, L., Kalther, J., Pomeroy, R., Abdullah, H., & Ruchimat, T. (2018). Review of national laws and regulation in Indonesia in relation to an ecosystem approach to fisheries management. *Marine Policy, 91,* 150–160. https://doi.org/10.1016/j.marpol.2018.01.027

Nurasa, T., Naamin, N., & Basuki, R. (1993). *The role of Panglima Laot 'sea commander' system in coastal fisheries management in Aceh, Indonesia.* Twenty-Second IPFC Fisheries Symposium, Darwin, Australia.

Organisation for Economic Co-operation and Development. (n.d.). *Country note on fisheries management systems—Australia*. OECD. Retrieved February 5, 2021, from https://www.oecd.org/australia/34427707.pdf

Ostrom, E. (1990). *Governing the commons: The evolution of institutions for collective action*. Cambridge University Press.

Patterson, H., Woodhams, J., Larcombe, J., & Curtotti, R. (2020). Chapter 1 overview. In H. Patterson, J. Larcombe, J. Woodhams, & R. Curtotti (Eds), *Fishery status reports 2020* (pp. 1–31). ABARES. Retrieved February 5, 2021, from https://www.agriculture.gov.au/abares/research-topics/fisheries/fishery-status/overview

Pikitch, E. K., Santora, C., Babcock, E. A., Bakun, A., Bonfil, R., Conover, D. O., Sainsbury, K. J., et al. (2004). Ecosystem-based fishery management. *Science, 305*(5682), 346–347. https://doi.org/10.1126/science.1098222

Pinkerton, E. (2017). Hegemony and resistance: Disturbing patterns and hopeful signs in the impact of neoliberal policies on small-scale fisheries around the world. *Marine Policy, 80*(2016), 1–9. https://doi.org/10.1016/j.marpol.2016.11.012

Pinkerton, E., & Edwards, D. N. (2009). The elephant in the room: The hidden costs of leasing individual transferable fishing quotas. *Marine Policy, 33*(4), 707–713. https://doi.org/10.1016/j.marpol.2009.02.004

Pomeroy, R. S., & Berkes, F. (1997). Two to tango: The role of government in fisheries co-management. *Marine Policy, 21*(5), 465–480. https://doi.org/10.1016/S0308-597X(97)00017-1

Quimby, B., & Levine, A. (2018). Participation, power, and equity: Examining three key social dimensions of fisheries comanagement. *Sustainability, 10*(9), 3324. https://doi.org/10.3390/su10093324

Ratner, B. D., Oh, E. J. V., & Pomeroy, R. S. (2012). Navigating change: Second-generation challenges of small-scale fisheries co-management in the Philippines and Vietnam. *Journal of Environmental Management, 107*, 131–139. https://doi.org/10.1016/j.jenvman.2012.04.014

Ruddle, K., & Satria, A. (Eds.). (2010). *Managing coastal and inland waters: Pre-existing aquatic management systems in Southeast Asia*. Springer.

Satria, A., & Adhuri, D. S. (2010). Pre-existing fisheries management systems in Indonesia, focusing on Lombok and Maluku. In K. Ruddle & A. Satria (Eds.), *Managing coastal and inland waters: Pre-existing aquatic management systems in Southeast Asia* (pp. 31–55). Springer.

Schnierer, S., & Egan, H. (2015). *Indigenous cultural fishing and fisheries governance*. Research Project. FRDC 2012/216. Fisheries Research and Development Corporation.

Schurman, R. A. (1998). Tuna dreams: Resource nationalism and the Pacific Islands' tuna industry. *Development and Change, 29*(1), 107–136. https://doi.org/10.1111/1467-7660.00072

Segi, S. (2014). Protecting or pilfering? Neoliberal conservationist marine protected areas in the experience of coastal Granada, the Philippines. *Human Ecology, 42*(4), 565–575. https://doi.org/10.1007/s10745-014-9669-1

Shaw, S., Johnson, H., & Dressler, W. (2011). *Identifying, communicating and integrating social considerations into future management concerns in inshore commercial fisheries in coastal Queensland.* Research Project. FRDC 2008/073. Canberra and Queensland: Fisheries Research and Development Corporation, University of Queensland, and Queensland Seafood Industry Association.

Stacey, N. E., Karam, J., Meekan, M. G., Pickering, S., & Ninef, J. (2012). Prospects for whale shark conservation in eastern Indonesia through Bajo traditional ecological knowledge and community-based monitoring. *Conservation and Society, 10*(1), 63–75. https://doi.org/10.4103/0972-4923.92197

Steenbergen, D. J. (2016). Strategic customary village leadership in the context of marine conservation and development in Southeast Maluku, Indonesia. *Human Ecology, 44*(3), 311–327. https://doi.org/10.1007/s10745-016-9829-6

Stevens, R., Cartwright, I., & Neville, P. (2012). *Independent review of NSW commercial fisheries policy, management and administration.* Department of Primary Industries. Retrieved February 5, 2021, from https://www.dpi.nsw.gov.au/__data/assets/pdf_file/0004/631633/Independent-Comm-Fish-Review-Report-Mar2012.pdf

Sukarsih, Y., Zulbainarni, N., & Jahroh, S. (2019). The impact of the moratorium and transhipment policies on the tuna fisheries business in Bitung Indonesia. *International Journal of Scientific Technology and Research, 8*(4), 329–332.

Sunderlin, W. D., & Gorospe, M. L. G. (1997). Fishers' organizations and modes of co-management: The case of San Miguel Bay, Philippines. *Human Organization, 56*(3), 333–343. https://doi.org/10.17730/humo.56.3.457 6w844451k342t

Sunoko, R., & Huang, H.-W. (2014). Indonesia tuna fisheries development and future strategy. *Marine Policy, 43*, 174–183. https://doi.org/10.1016/j.marpol.2013.05.011

Sutton, M. (2020). South Australian reforms to put two-thirds of local commercial fishers out of business, stakeholders say. Retrieved December 2, 2020, from https://www.abc.net.au/news/2020-11-10/south-australian-scalefish-fishery-reforms-impact-on-fishers/12861520

Thorburn, C. C. (2000). Changing customary marine resource management practice and institutions: The case of Sasi Lola in the Kei Islands, Indonesia. *World Development, 28*(8), 1461–1479. https://doi.org/10.1016/S0305-750X(00)00039-5

Voyer, M., Barclay, K., McIlgorm, A., & Mazur, N. (2016). *Social and economic evaluation of NSW coastal professional wild-catch fisheries.* Research Project. FRDC 2014/301. University of Technology Sydney.

Voyer, M., Barclay, K., McIlgorm, A., & Mazur, N. (2017). Connections or conflict? A social and economic analysis of the interconnections between the professional fishing industry, recreational fishing and marine tourism in coastal communities in NSW, Australia. *Marine Policy, 76,* 114–121. https://doi.org/10.1016/j.marpol.2016.11.029

Voyer, M., Quirk, G., McIlgorm, A., & Azmi, K. (2018). Shades of blue: What do competing interpretations of the blue economy mean for oceans governance? *Journal of Environmental Policy & Planning, 20*(5), 595–616. https://doi.org/10.1080/1523908X.2018.1473153

Wallace-Carter, E. (1987). *For they were fishers: The history of the fishing industry in South Australia.* Amphitrite.

Wilkinson, J. (1997). *Commercial fishing in NSW origins and development to the 1990s.* Briefing Paper no. 15/1997. NSW parliamentary library research service. Retrieved February 5, 2021, from https://www.parliament.nsw.gov.au/researchpapers/Pages/commercial-fishing-in-nsw-origins-and-developmen.aspx

Wilson, C., & Linkie, M. (2012). The Panglima Laot of Aceh: A case study in large-scale community-based marine management after the 2004 Indian Ocean tsunami. *Oryx, 46*(4), 495–500. https://doi.org/10.1017/S0030605312000191

World Bank. (2015). *Indonesia systematic country diagnostic: Connecting the bottom 40 percent to the prosperity generation.* Report no. 94066-ID. World Bank.

Zerner, C. (1994). Through a green lens: The construction of customary environmental law and community in Indonesia's Maluku Islands. *Law and Society Review, 28*(5), 1079–1122. https://doi.org/10.2307/3054024

CHAPTER 5

Fishing Livelihoods and Wellbeing

Abstract The final chapter of this book discusses the implications of a relational approach to fishing livelihoods for governance for improved social and ecological outcomes. The chapter reviews some of the ways in which academics, activists and policymakers can use approaches that emphasise the relational context of fishing livelihoods, and specifies the concept of wellbeing as one that can usefully and practically build bridges between fisheries stakeholders with diverse interests. The chapter then examines two assessments of fisheries on community wellbeing: the social and economic impacts of fisheries in Australia, and the effects of governance on wellbeing of fishing communities in Indonesia and Solomon Islands.

Keywords Fisheries governance • Wellbeing • Relationality

Fishing livelihoods, especially in the Asia-Pacific, remain hugely significant. They produce healthy and nutritious food (Hicks et al., 2019), generate economic opportunities for many millions, play a particularly important role for vulnerable and marginalised groups (Mills et al., 2011) and contribute to the maintenance of traditions and cultures (Allison et al., 2020). Yet, the environmental crises progressively enveloping the globe are particularly acute in the marine systems on which fishing

M. Fabinyi, K. Barclay, *Asia-Pacific Fishing Livelihoods*,
https://doi.org/10.1007/978-3-030-79591-7_5

livelihoods depend. Climate change, pollution and overfishing are among the many drivers of change to these marine systems that threaten their capacity to sustainably generate marine resources. At the same time, economic, political and social drivers of change are reshaping the structures of social life that dictate how fishing livelihoods operate. While fishing livelihoods have evolved and adapted to many changes over the years, the accelerating scale and pace of change present significant challenges to the very viability of fishing livelihoods in some places. The Asia-Pacific, home to the largest number of fishers and the most diverse marine ecosystems on the planet, is a crucial locus of these developments (Fig. 5.1).

In much fisheries governance literature, fishers are represented as individuals whose sole objective is to maximise the number of fish that they catch, with subsequent environmental effects. From this perspective, governance is a balancing act that seeks to maximise the acquisition of financial wealth while minimising environmental harms. In contrast, this short book has tried to highlight some of the relationships that drive the perspectives and actions of those working in fisheries value chains. Fishing does not occur in isolation, but takes place in relation to a wider environment of other activities, actors and ideas. We have highlighted three relationships that we argue are particularly important to fishing livelihoods: historical patterns of economic and political change, social identities and relations, and institutional structures. Each, in its own way, is an important part of the nature and character of particular fishing livelihoods and, thus, contributes to the diverse social and ecological outcomes associated with fishing.

While these relationships are usually studied from different conceptual perspectives, or in relative isolation from each other, they are largely complementary (Hornborg et al., 2013). In the Philippines, for example, the cases in this book showed how contemporary class structures in fishing livelihoods derive from historical patterns of political and economic change, intertwine with cultural values relating to inequality and influence the differentiated outcomes of governance interventions. Some of the key concepts of political ecology introduced in Chap. 1 are useful reminders of the ways in which fishing livelihoods are constituted by these multiple, shifting relationships. Attention to wider scales of analysis and to historical pathways of change, for example, shows how the conditions of fishing livelihoods are generated by broader processes that go well beyond the day-to-day activities of harvesting fish and other local activities. And recognition that politicised environments (Bryant & Bailey, 1997) are the

Fig. 5.1 A fisher returns from a night at sea in North-East Palawan, Philippines. (Photo credit: Katherine Jack)

norm rather than unusual shows how the social relationships constituting fishing livelihoods, and the governance interventions that seek to manage them, are experienced in unequal and distinctive ways by different individuals and groups.

This concluding chapter sets out the pragmatic implications of this relational perspective in the applied sphere. When the focus shifts from the act

of fishing itself to the wider sets of relationships in which fishing livelihoods are embedded, opportunities for action emerge in new spaces.
Understanding fishing in relation to past and present patterns of economic
development, in relation to social inequities, and in relation to the social
effects of governance regimes, increasingly informs the work of many academics, activists and even those in government. The fundamental role of
markets and trade in fishing has led to the formation of new coalitions
seeking to harness the power of these markets for improved, more ecologically sustainable growth through the sustainable seafood movement. On
the other side of the ideological spectrum, focusing on unequal patterns
of economic growth has led to trenchant critiques of the current economic system and its incarnations in 'blue growth' policies (Mallin &
Barbesgaard, 2020) or industrial fishing (Longo et al., 2015), as well as
advocacy of alternatives such as degrowth (Hadjimichael, 2018) and blue
justice (Isaacs, 2019). Similarly, environmentalists draw attention to the
enormous costs of our economic system for the environment, in particular
for carbon emissions and climate change (Hughes et al., 2017).

At a more micro scale, understanding how patterns of power reproduce
themselves in day-to-day interactions and relationships informs much academic and applied social science work on fishing livelihoods. For example,
much work on value chains seeks to improve the capacity of fishers to
'upgrade' their position in the value chain, through the development of
new skills or access to new technologies (Cole et al., 2018; Purcell, 2014),
or access to better market information (Purcell et al., 2017). Increasingly,
governments, donors and environmental NGOs are working to address
the position of marginalised groups such as women (Barclay et al., 2019;
Kleiber et al., 2019; Lawless et al., 2017; USAID Oceans and Fisheries
Partnership, 2019). The issue of working conditions and rights among
fishers, particularly those working on industrial fishing vessels, is now
prompting governments and corporations to address these issues (Kittinger
et al., 2017).

Attention to the relationships between fishing livelihoods and institutions, particularly institutional inequities, also informs much social science
work and activism, and has driven change in dominant models of fisheries
governance. Attention to the role and rights of communities was one of
the major drivers behind co-management, for example, which has been
adopted in various ways in many parts of the Asia-Pacific (Ratner et al.,
2012). Similarly, in many contexts state models of fisheries governance
must work with customary marine tenure and access rules (Rohe et al.,

2019). A focus on the rights of fishers has led to the development of the Voluntary Guidelines for Securing Sustainable Small-Scale Fisheries, and a thriving coalition of groups advocating for greater visibility and support for small-scale fishers (Jentoft, 2019).

In the remainder of this chapter, we return to the theme of wellbeing introduced in Chap. 1 and discuss some of the practical ways in which this concept can be operationalised and used to contribute towards better ecological and social outcomes. The concept of wellbeing is no panacea (Young et al., 2018) or blueprint (Ratner & Allison, 2012); it is not a new model for governance that can be expected to work effectively in the same way in every context. However, one of the strengths of the wellbeing concept is its ability to bridge between different stakeholders involved in fisheries governance and scholarship.

For many fisheries policymakers and managers, the findings of social scientists (e.g., detailed ethnographic investigations) are often 'interesting' but ultimately difficult to operationalise, 'unscientific', or 'a bit touchy feely', as one prominent fisheries scientist advised author Fabinyi at a technical workshop. Conversely, for many social scientists, such as anthropologists fascinated by the complexities and contradictions in human societies, the tendency in many economic models and in much fisheries policy to discuss fishers in terms of numbers (catch, value, volume, number of fishers, etc.) is a gross simplification of underlying key social processes and structures. So often, the debates between economists, policymakers and social scientists become 'bogged down' in sterile and unproductive arguments because of these fundamentally different assumptions about knowledge. Wellbeing, we argue, is a concept that can help to incorporate some of the complex issues addressed by social scientists in a way that can be recognised, understood and acted on by fisheries policymakers. It is useful for exploring the interrelated environmental, political and economic aspects of fisheries.

One reason the wellbeing approach is useful for bringing together disparate knowledge systems is that it addresses a shared high-level goal: the wellbeing of human communities (Stiglitz et al., 2018). Using fisheries resources for the benefit of the people is usually the overarching aim stated in fisheries legislation, so it is clearly in the purview of fisheries managers. The conceptualisation of wellbeing most often used in fisheries draws from development studies, adapted from Amartya Sen's capabilities approach (Sen et al., 1987). It is described as 'a state of being with others, which arises where *human needs are met*, where one can act meaningfully to pursue one's goals, and where one can enjoy a satisfactory quality of life' (McGregor 2008, cited in Coulthard et al., 2011: 79).

Another reason the wellbeing approach enables participation from across disciplines is that it is a framework, rather than a method itself. Various methods are used, often mixed to gain a comprehensive picture (McGregor et al., 2015), such as qualitative interviews, social psychological tools (Britton & Coulthard, 2013; Coulthard et al., 2014), semi-quantitative questionnaires, economic analysis (Voyer et al., 2017) and ecosystem services methods (Chaigneau et al., 2019; Masterson et al., 2019). Thus, people from different disciplinary perspectives are usually able to see some kind of method they recognise as rigorous and appropriate in wellbeing studies. The holistic emphasis in wellbeing studies means that different domains of life are examined—from straightforward economic benefits to political relationships.

Wellbeing has been used to assess the condition of fishing communities (Britton & Coulthard, 2013; Coulthard et al., 2011, 2014; Smith & Clay, 2010) as well as the human dimensions in ecosystem-based resource management (e.g., Breslow et al., 2016). These uses of the wellbeing approach involve asking the question, 'what is the wellbeing of community x?' For example, the University of Canberra runs annual wellbeing surveys of rural areas in Australia, using established psychological questionnaires for individual and community wellbeing, enabling comparison across regions, across demographic groups in populations, and over time.[1] It has been used to compare the wellbeing of fishers against the rest of the population (Barclay et al., 2020).

Wellbeing has also been used to assess effects on communities. These uses involve asking the question, 'what are the effects of y on the wellbeing of community x?' (e.g., measuring the wellbeing benefits people gain from coastal ecosystem services) (Chaigneau et al., 2019; Masterson et al., 2019; McMichael et al., 2005). Thus, wellbeing is a suitable framework for social impact assessments. Other conceptual frameworks used for understanding social impacts in fishing communities include resilience and vulnerability. There are some examples of ongoing monitoring of social and economic conditions in relation to marine ecosystems; although, these do not use a wellbeing framework (i.e., the Social and Economic

---

[1] Information about the Australian Regional Wellbeing Survey is available at https://www.canberra.edu.au/research/institutes/health-research-institute/regional-wellbeing-survey/survey-results

Long-Term Monitoring Program for the Great Barrier Reef in Australia,[2] and the collection of data on social indicators for coastal and ocean ecosystems across the US) (Ramenzoni & Yoskowitz, 2017). A study in Bangladesh showed the effects of different types of aquaculture through comparing the wellbeing of a village engaged in rearing tiger shrimp with another village rearing freshwater prawn (Belton, 2016). The wellbeing approach has also been used to specifically assess the effects of fisheries on community wellbeing. The following cases examine two such assessments.

SOCIAL AND ECONOMIC EFFECTS OF FISHERIES IN AUSTRALIA

There have been two large evaluations of the effects of fisheries on the wellbeing of communities in Australia, for the states of New South Wales (Voyer et al., 2016, 2017) and Victoria (Abernethy et al., 2020). The impetus for these studies was the lack of public and government support for fisheries in Australia noted in Chap. 4. Fishing industry bodies felt that robust evidence about the positive effects fisheries have on communities where fishing occurs would help their advocacy efforts to improve public perceptions about fisheries, and ensure continued access to fisheries resources.

The studies involved a two-step process. First, the areas of community life or domains of wellbeing to which fisheries can contribute were identified, through the literature on wellbeing and qualitative interviews with fishers and others in fishing communities. Second, the specific contributions fisheries can make to those areas of community life were identified (see Table 5.1). These contributions were investigated through a mix of qualitative interviews, document review, a semi-quantitative phone survey, and economics analysis, including contributions to regional economies using input–output methods.

Since wellbeing is multidimensional (McGregor et al., 2015; Stiglitz et al., 2018), fisheries contributions were considered as having material, subjective and relational dimensions. Material contributions are easy to understand—they include food, income and assets, access to services, and environmental quality. Fisheries contributions to subjective wellbeing are effects on people's perceptions of their quality of life and the values and

---

[2] The Social and Economic Long-Term Monitoring Program data for the Great Barrier Reef are available at https://data.csiro.au/dap/landingpage?pid=csiro:38797

Table 5.1 Fisheries-relevant domains of community wellbeing, and fisheries contributions to those domains

Domains of community wellbeing				
Economic diversity and resilience	Food supply	Tourism and recreation	Environmental sustainability	Social fabric of communities
Contributions of fisheries				
Revenue Employment Synergies with connected industries: Service, post-harvest, tourism	Fresh local seafood Nutritious food Food safety	Local seafood for visitors Experiences and aesthetics for visitors Supporting other activities, including recreational fishing	Fishery monitoring and research Improving sustainable fishing practices Participating in environmental research and stewardship activities	Local sense of place and identity as 'fishing town' Supporting community life: Donations, volunteering Workplace for vulnerable young men

Source: Abernethy et al. (2020)

beliefs that shape their levels of satisfaction, such as whether they feel it is a good thing to be eating locally produced seafood or believe that fishers are operating in ways that sustain the marine environment. Relational aspects include whether fishers contribute to the development and maintenance of relationships that enable communities to achieve wellbeing, such as through donating to or volunteering in community activities like sports or festivals, or through business and political connections that may benefit communities.

The picture of fisheries that emerges from a wellbeing analysis really illuminates livelihoods. Before these studies were conducted the only existing data were statewide fisheries' gross value of production (i.e., volume of catch multiplied by beach price), and rough job numbers from census data. The wellbeing economic analysis expanded the view to consider businesses supplying services and gear to fishers, as well as the flow-on to businesses in seafood processing and wholesaling. The input–output method estimated the level of economic activity fishing generated in regions, down to local government areas, as well as the proportion of this going to household incomes, and the numbers of jobs in fishing, processing and wholesaling.

This kind of material information is certainly useful for making claims about the importance of fishing to communities, but the multidimensional mixed-methods approach greatly deepened the understanding generated. For example, the qualitative interviews and phone survey revealed the close connections between fisheries and tourism—another key economic activity in many fishing communities. It is not just that the two sectors support each other, but tourism in many locations is seasonal whereas fishing is year round, so having fishing as well as tourism is important for local economies. The interviews also revealed that it is not only the numbers of jobs that are important but also the types. Entry-level employment in fisheries is valued in rural areas, because without entry-level work young people have to leave to seek employment elsewhere. Moreover, fisheries work does not require high levels of formal schooling, so it has been a good opportunity for young men who struggled at school or who were 'getting into trouble' and likely heading towards a life of crime and/or welfare dependency.

Finally, looking for contributions to subjective aspects of fishing reveals another element about livelihoods. Some fishers are deeply attached to fishing as a way of life. When they are forced to leave fishing work they become depressed. For these people working on the water, feeling the majesty of the elements and other living creatures, and the satisfaction from surviving and thriving in this work, is about much more than the income they make: it is part of their psychological wellbeing. There is an added layer of importance for Indigenous peoples. Australian Aboriginal and Torres Strait Islanders gain significant and measurable benefits from working on Country, caring for their environment, gaining sustenance from it, sharing food from it and developing and sharing knowledge about it. Therefore, having the opportunity to pursue a livelihood in fishing is about far more than cash income or dietary nutrients—although those material elements are also undeniably important. A wellbeing approach examining relational and subjective, as well as material dimensions, gives a holistic picture of the effects fishing has on communities.

Effects of Governance on Wellbeing of Fishing Communities in Indonesia and Solomon Islands

A slightly different tack was taken in looking at the wellbeing contributions of tuna fisheries to coastal communities in Indonesia and Solomon Islands (McClean et al., 2019). The impetus for this study was a deficit in existing fisheries governance analyses. Around the world the overarching objective for fisheries management enshrined in fisheries legislation is almost always to benefit society, but fisheries management science virtually never considers the social or economic benefits from fisheries in a systematic way: fisheries management science is usually about fish stocks. Sometimes there is an economic element such as 'maximum economic yield' or bio-economic modelling—both of which often focus on fishing business profitability—or more basic information about contribution to national GDP, gross value of production or total numbers of jobs. Each of these approaches is 'broad brush' and does not give the fine-grained, multidimensional understanding of economic effects achieved in the Australian studies using a wellbeing approach. Non-economic social benefits are rarely examined in fisheries science at all.

Therefore, the project asked the question, 'what are the effects of different types of fisheries governance on the wellbeing generated through tuna fisheries in coastal communities?' The stages involved in answering this question included identifying (1) the fisheries-relevant domains of wellbeing in Indonesia and Solomon Islands contexts; (2) the contributions tuna fisheries make to those domains; (3) the types of governance affecting those contributions; and (4) the effects those forms of governance have on wellbeing contributions from tuna fisheries. 'Governance' here is understood broadly to include all social, economic and political institutions shaping fisheries, as well as specific fisheries regulations, as per insights from governance studies (Kooiman et al., 2005). For example, private sector interventions such as certification for MSC and fair trade and differential access to markets are part of governance.

The scope of this project precluded economic analysis or large-scale questionnaire data collection. It relied on qualitative interviews (134 across both countries) and review of documents, and existing social, economic and fisheries data. It would be useful to also have quantitative social and economic data and analysis. At the time of writing the researchers were pursuing this in subsequent research.

However, even without quantitative data or analysis the qualitative wellbeing analysis produced an approach for assessing governance that can help fisheries managers and stakeholders start to grasp the implications of different governance interventions for the wellbeing of target communities. Table 5.2 outlines a framework that can be used to conduct a 'first pass' assessment that can help illuminate likely social and economic effects of a governance intervention. This framework can also highlight

Table 5.2 Framework for assessing fisheries governance in terms of community wellbeing

The potential governance intervention	List the intended changes in a fishery, or set of fisheries management options
The fishery affected	Include relevant information on gear/vessel type, target species, geographical focus, destination market or any other characteristics of the fishery that are relevant in determining the scope of the intervention.
Status quo wellbeing impacts	List the actors, communities or stakeholders who currently receive benefits, or are exposed to risk or insecurity, from the fishery. Wellbeing impacts may be in the domains of economy, food and nutritional security, workplace health and safety, healthy environmental systems and poverty alleviation. Close consideration should be paid to the distribution of those benefits according to socio-economic status, migrant status and gender.
Potential benefits to coastal communities from the intervention	Note intended or anticipated wellbeing benefits that would arise from the initiative, as well as whether these are likely to be realised in the short, medium or long term.
Who benefits from the governance intervention?	List the actors, communities or stakeholders who would receive the benefit. Close consideration should be paid to socio-economic status, participation of migrant communities or migrant labour, and gender.
Potential lost benefits to coastal communities	Note wellbeing benefits that may be lost as a result of the intervention (such as livelihoods if catches are restricted), with likely time frame (short, medium or long term).
Who in the value chain bears the loss/ is exposed to risk from the intervention	List the actors, communities or stakeholders who might lose benefits, or be exposed to risks. Close consideration should be paid to socio-economic status, migrant status and gender.
Factors influencing effectiveness and the ability to mitigate risks/vulnerabilities	List any factors likely to influence the effectiveness of an initiative, or mitigate the risks of an initiative (e.g., the presence of alternative livelihoods and food sources, or the presence of effective monitoring or management systems). This allows for realistic assessment of the feasibility of an initiative in the context of a specific fishery and management system.

Source: McClean et al. (2019)

knowledge gaps, both to orient future research efforts to provide better data and quantitative analysis, and improve the evidence base. For example, this includes household incomes coming from tuna industries, numbers and types of jobs, which social groups have which types of jobs, and contributions to regional economies, among other factors. Ideally, there should be ongoing monitoring of social and economic aspects of fisheries, so that social and economic factors can become part of routine fisheries science, and so that data are available for when social and economic impact assessments are needed.

The main areas of wellbeing benefit that emerged in the project were economy and livelihoods, food and nutritional security, and environmentally sustainable fisheries. Tuna industries give rise to formal and informal employment and large and small business opportunities in fishing, trading and processing, as well as in supplying inputs and services for fishing and processing. Indonesian Government statistics on fisheries employment do not break down the numbers for tuna, but it is a prominent part of the economy in many coastal areas. For example, in the tuna port of Bitung in North Sulawesi, in 2018 there were over 8000 people employed in fishing and processing, with many more working in supply chains for markets in the regions surrounding Bitung (McClean et al., 2019). In Solomon Islands the tuna fishing and processing sector based in Noro in Western Province has long been the largest private sector employer in the country. It formally employed around 2400 people in 2018, and has long been a large employer of women, who work on the processing lines (McClean et al., 2019). In addition, there are hundreds of people doing small-scale tuna fishing and selling tuna in urban markets, both catch from the small-scale fisheries and rejects from the industrial fisheries. The cannery in Noro, which has a canteen for its workers, buys fresh vegetables from around 500 farmers in the surrounding area (McClean et al., 2019).

The subjective aspects of livelihoods were given less weight by interviewees than in the Australian studies, possibly due to the different situations of Indonesia and Solomon Islands as developing countries. Interviewees were mainly concerned with the material dimensions of livelihoods. Nevertheless, the broad view of the wellbeing approach—paying attention to relational and subjective dimensions, as well as material and the use of qualitative methods—helped develop a holistic picture of livelihoods. The two key findings about tuna livelihoods that emerged from this study were the workplace health and safety and income insecurity risks that are involved in some livelihoods, and the distribution of different livelihood opportunities across groups within society, particularly in relation to socio-economic status, ethnicity and/or migrant status, and gender.

The quality of many tuna livelihoods in Indonesia and Solomon Islands is greatly affected by security of income as well as workplace health and safety. Indonesian fishing crews are largely informally recruited: they have no contract, insecure catch-share models of remuneration and little health insurance coverage. There are injuries working with heavy equipment at sea, and crews of small-scale fishing vessels are sometimes lost. Indonesian tuna fishing crews are often among those found to be suffering human rights abuses through poor labour conditions, and even forced labour. In contrast, formal sector workers in processing factories are contracted, have minimum wage conditions, and are usually covered by various forms of social and health insurance. In Solomon Islands working conditions on vessels in the domestic fleet are some of the best in the region. Formal tuna processing work is similar in conditions and protections to Indonesia. Informal fishing, processing and trading work are likewise more insecure, without insurance protection, and small-scale fishing is similarly dangerous, with crew occasionally lost at sea.

The incomes for entry-level work in both Indonesia and Solomon Islands, in both formal and informal sectors, are very low, often around the poverty line—although this is not always the case. For example, Solomon Islands women cooking and selling tuna informally in markets make a great deal more than cannery workers. Financial literacy also makes a big difference in Solomon Islands, with some small-scale fishers and cannery workers who have had financial literacy training able to make their incomes go a lot further than others in terms of housing, covering weekly household expenses and so on. In Indonesia, some petty traders have worked their way up into running lucrative businesses (McClean et al., 2019).

Thus, entry-level tuna industry work in Indonesia is mainly filled by low socio-economic status groups. Often, this corresponds with migrant status, which can mean internal migration within Indonesia, such as Butonese fishers operating in Maluku. In Bitung there are many different internal migrants drawn by the availability of tuna work, and there has historically been a strong presence of Filipino fishers in Bitung. The Butonese people fishing in Maluku are marginalised and tuna fishing is one of the only options available to them. In Solomon Islands remuneration in the tuna sector is on par with other sectors, so formal tuna work is sought by people from all groups in society. However, small-scale tuna fishing does align with low socio-economic or migrant status. The reasons for this are complicated and include the fact that people with I-Kiribati heritage have more skills and knowledge in offshore fishing than most Indigenous Solomon Islanders, but according to interviewees also include the marginalised status of people who migrated from Kiribati. More research would be needed to observe whether tuna livelihoods act to alleviate poverty, or if the low incomes keep marginalised people in low socio-economic situations.

Another clear social differentiation in tuna livelihoods is by gender. Tuna fisheries are some of the more male-dominated fisheries, and very few women are engaged in tuna fishing in Solomon Islands or Indonesia. However, women are heavily involved in processing and trading. In formal processing women in both countries make up the majority of workers on tuna 'cleaning' lines, preparing the tuna meat for putting in cans. They work in administration, management and technical roles in processing factories. Women as well as men are also involved in informal trade of tuna and small-scale processing such as smoking or cooking tuna for sale. However, in both countries, men tend to cluster around the larger, higher-value parts of business, with women correspondingly clustered around smaller-scale, lower-value activities and lower-authority positions. Therefore, livelihood opportunities in tuna industries in both countries are shaped by gender as well as ethnicity and socio-economic status.

CONCLUSION

Fishing livelihoods currently face a series of profound and interrelated economic, environmental and social challenges that negatively affect and threaten their viability. While these challenges are increasingly well understood—from climate change to overfishing, and from new government regulation to increasing competition over marine and inland fisheries resources with other economic sectors—the solutions to these challenges are much harder to encounter. This reflects the reality that fisheries and coastal governance is inherently a 'wicked problem'—one where there are many dynamic factors and stakeholders, where there are no technical solutions and where it is not even clear when the problem is 'solved' (Jentoft & Chuenpagdee, 2009).

The final chapter of this book discussed how the wellbeing approach can help to shift the dominant focus in fisheries governance and policy-making from economic benefits and environmental effects alone, to one that considers the wider relations by which fishing livelihoods are shaped. A wellbeing approach as described in this chapter can reveal the different dimensions of wellbeing to which fishing livelihoods contribute, and how fisheries managers and stakeholders can assess the consequences of specific governance interventions for the wellbeing of fishing communities.

Fishing livelihoods will continue to change in relation to the wider world of which they are part. Understanding and incorporating these relationships into the knowledge informing decision-making is one way that the wider fisheries policymaking and management community can ultimately contribute to the improvement and sustainability of fishing livelihoods.

REFERENCES

Abernethy, K., Barclay, K., McIlgorm, A., Gilmour, P., McClean, N., & Davey, J. (2020). *Victoria's fisheries and aquaculture: Economic and social contributions*. Research project. FRDC 2017-092. Sydney: University of Technology Sydney. Retrieved January 7, 2021, from https://www.uts.edu.au/about/faculty-arts-and-social-sciences/research/fass-research-projects/social-science-fisheries/victorias-fisheries-and-aquaculture-economic-and-social-contributions

Allison, E.H., Kurien, J., Ota, Y., Adhuri, D. S., Bavinck, J.M., Cisneros-Montemayor, A., Olukoju, A., et al. (2020). *The human relationship with our ocean planet*. Washington, DC: World Resources Institute. Retrieved February 9, 2021, from https://www.oceanpanel.org/blue-papers/HumanRelationshipwithOurOceanPlanet

Barclay, K., Leduc, B., Mangubhai, S., & Donato-Hunt, C. (Eds.). (2019). *Pacific handbook for gender equity and social inclusion in coastal fisheries and aquaculture* (1st ed.). Noumea.

Barclay, K., Davila, F., Kim, Y., McClean, N., & Mcilgorm, A. (2020). *Economic analysis and social and economic monitoring following the NSW Commercial Fisheries Business Adjustment Program*. Sydney: Institute for Sustainable Futures, University of Technology Sydney. Retrieved January 7, 2021, from https://www.dpi.nsw.gov.au/__data/assets/pdf_file/0007/1256128/Economic-analysis-and-Social-and-Economic-monitoring-following-the-NSW-Commercial-Fisheries-Business-Adjustment-Program.pdf

Belton, B. (2016). Shrimp, prawn and the political economy of social wellbeing in rural Bangladesh. *Journal of Rural Studies, 45*, 230–242. https://doi.org/10.1016/j.jrurstud.2016.03.014

Breslow, S. J., Sojka, B., Barnea, R., Basurto, X., Carothers, C., Charnley, S., Coulthard, S., et al. (2016). Conceptualizing and operationalizing human wellbeing for ecosystem assessment and management. *Environmental Science and Policy, 66*, 250–259. https://doi.org/10.1016/j.envsci.2016.06.023

Britton, E., & Coulthard, S. (2013). Assessing the social wellbeing of Northern Ireland's fishing society using a three-dimensional approach. *Marine Policy, 37*, 28–36. https://doi.org/10.1016/j.marpol.2012.04.011

Bryant, R. L., & Bailey, S. (1997). *Third world political ecology*. Routledge.

Chaigneau, T., Brown, K., Coulthard, S., Daw, T. M., & Szaboova, L. (2019). Money, use and experience: Identifying the mechanisms through which ecosystem services contribute to wellbeing in coastal Kenya and Mozambique. *Ecosystem Services, 38,* 100957. https://doi.org/10.1016/j.ecoser.2019.100957

Cole, S. M. C., McDougall, C., Kaminski, A. M., Kefi, A. S., Chilala, A., & Chisule, G. (2018). Postharvest fish losses and unequal gender relations: Drivers of the social-ecological trap in the Barotse Floodplain fishery, Zambia. *Ecology and Society, 23*(2), 18. https://doi.org/10.5751/ES-09950-230218

Coulthard, S., Johnson, D., & McGregor, J. A. (2011). Poverty, sustainability and human wellbeing: A social wellbeing approach to the global fisheries crisis. *Global Environmental Change, 21*(2), 453–463. https://doi.org/10.1016/j.gloenvcha.2011.01.003

Coulthard, S., Sandaruwan, L., Paranamana, N., & Koralgama, D. (2014). Taking a well-being approach to fisheries research: Insights from a Sri Lankan fishing village and relevance for sustainable fisheries. In L. Camfield (Ed.), *Methodological challenges and new approaches to research in international development* (pp. 76–100). Palgrave Macmillan.

Hadjimichael, M. (2018). A call for a blue degrowth: Unravelling the European Union's fisheries and maritime policies. *Marine Policy, 94,* 158–164. https://doi.org/10.1016/j.marpol.2018.05.007

Hicks, C. C., Cohen, P. J., Graham, N. A. J., Nash, K. L., Allison, E. H., D'Lima, C., MacNeil, M. A., et al. (2019). Harnessing global fisheries to tackle micronutrient deficiencies. *Nature, 574*(7776), 95–98. https://doi.org/10.1038/s41586-019-1592-6

Hornborg, A., Clark, B., & Hermele, K. (2013). Introduction: Ecology and power. In A. Hornborg, B. Clark, & K. Hermele (Eds.), *Ecology and power: Struggles over land and material resources in the past, present and future* (Vol. 18). Routledge.

Hughes, T. P., Barnes, M. L., Bellwood, D. R., Cinner, J. E., Cumming, G. S., Jackson, J. B., Palumbi, S. R., et al. (2017). Coral reefs in the Anthropocene. *Nature, 546*(7656), 82–90. https://doi.org/10.1038/nature22901

Isaacs, M. (2019). *Is the blue justice concept a human rights agenda?* University of the Western Cape, Institute for Poverty, Land and Agrarian Studies Policy Brief No. 54. Retrieved February 9, 2021, from https://www.africaportal.org/documents/20620/POLICY_BRIEF_54-BLUE_JUSTICE.pdf

Jentoft, S. (2019). *Life above water: Essays on human experiences of small-scale fisheries.* TBTI Global Book Series 1. TBTI Global.

Jentoft, S., & Chuenpagdee, R. (2009). Fisheries and coastal governance as a wicked problem. *Marine Policy, 33*(4), 553–560. https://doi.org/10.1016/j.marpol.2008.12.002

Kittinger, J. N., Teh, L. C. L., Allison, E. H., Bennett, N. J., Crowder, L. B., Finkbeiner, E. M., Young, J., et al. (2017). Committing to socially responsible seafood. *Science, 356*(6341), 912–913. https://doi.org/10.1126/science.aam9969

Kleiber, D., Cohen, P., Gomese, C., & McDougall, C. (2019). *Gender-integrated research for development in Pacific coastal fisheries.* Program Brief: FISH-2019-02. Penang: CGIAR Research Program on Fish Agri-Food Systems.

Kooiman, J., Bavinck, M., Jentoft, S., & Pullin, R. (Eds.). (2005). *Fish for life: Interactive governance for fisheries.* Amsterdam University Press.

Lawless, S., Doyle, K., Cohen, P., Eriksson, H., Schwarz, A.M., Teioli, H., Vavekaramui, A., et al. (2017). Considering gender: Practical guidance for rural development initiatives in Solomon Islands. Program Brief 2017-22. Penang: WorldFish.

Longo, S. B., Clausen, R., & Clark, B. (2015). *The tragedy of the commodity: Oceans, fisheries, and aquaculture.* Rutgers University Press.

Mallin, F., & Barbesgaard, M. (2020). Awash with contradiction: Capital, ocean space and the logics of the blue economy paradigm. *Geoforum, 113,* 121–132. https://doi.org/10.1016/j.geoforum.2020.04.021

Masterson, V. A., Vetter, S., Chaigneau, T., Daw, T. M., Selomane, O., Hamann, M., Wong, G. Y., et al. (2019). Revisiting the relationships between human well-being and ecosystems in dynamic social-ecological systems: Implications for stewardship and development. *Global Sustainability, 2*(e8), 1–14. https://doi.org/10.1017/S205947981900005X

McClean, N., Barclay, K., Fabinyi, M., Adhuri, D.S., Sulu, R.J., Indrabudi, T. (2019). Assessing tuna fisheries governance for community wellbeing: Case studies from Indonesia and Solomon Islands. Sydney: University of Technology Sydney. Retrieved January 7, 2021, from https://www.uts.edu.au/about/faculty-arts-and-social-sciences/research/fass-research-projects/assessing-governance-tuna

McGregor, A., Coulthard, S., & Camfield, L. A. (2015). *Measuring what matters: The role of well-being methods in development policy and practice.* Development Progress Project Note 04. London: Overseas Development Institute.

McMichael, A., Scholes, R., Hefny, M., Pereira, E., Palm, C., & Foale, S. (2005). Linking ecosystem services and human well-being. In D. Capistrano, K. C. Samper, M. J. Lee, & C. Raudsepp-Hearne (Eds.), *Ecosystems and human well-being: Multi-scale assessments* (pp. 43–60). Millennium Ecosystem Assessment Series 4. Washington, DC: Island Press.

Mills, D. J., Westlund, L., de Graaf, G., Kura, Y., Willman, R., & Kelleher, K. (2011). Under-reported and undervalued: Small-scale fisheries in the developing world. In R. S. Pomeroy & N. L. Andrew (Eds.), *Small-scale fisheries management: Frameworks and approaches for the developing world* (pp. 1–15). Cabi.

Purcell, S.W. (2014). *Processing sea cucumbers into bêche-de-mer: A manual for Pacific Island fishers*. Noumea: Secretariat of the Pacific community; Lismore: Southern Cross University.

Purcell, S. W., Crona, B. I., Lalavanua, W., & Eriksson, H. (2017). Distribution of economic returns in small-scale fisheries for international markets: A value-chain analysis. *Marine Policy, 86,* 9–16. https://doi.org/10.1016/j.marpol.2017.09.001

Ramenzoni, V. C., & Yoskowitz, D. (2017). Systematic review of recent social indicator efforts in US coastal and ocean ecosystems (2000–2016). *Environment and Society: Advances in Research, 8*(1), 9–39. https://doi.org/10.3167/ares.2017.080102

Ratner, B. D., & Allison, E. H. (2012). Wealth, rights, and resilience: An agenda for governance reform in small-scale fisheries. *Development Policy Review, 30*(4), 371–398. https://doi.org/10.1111/j.1467-7679.2012.00581.x

Ratner, B. D., Oh, E. J. V., & Pomeroy, R. S. (2012). Navigating change: Second-generation challenges of small-scale fisheries co-management in the Philippines and Vietnam. *Journal of Environmental Management, 107,* 131–139. https://doi.org/10.1016/j.jenvman.2012.04.014

Rohe, J. R., Govan, H., Schlüter, A., & Ferse, S. C. A. (2019). A legal pluralism perspective on coastal fisheries governance in two Pacific Island countries. *Marine Policy, 100,* 90–97. https://doi.org/10.1016/j.marpol.2018.11.020

Sen, A., Muellbauer, J., & Hawthorn, G. (1987). *The standard of living*. Cambridge University Press.

Smith, C. L., & Clay, P. M. (2010). Measuring subjective and objective well-being: Analyses from five marine commercial fisheries. *Human Organization, 69*(2), 158–168. https://doi.org/10.17730/humo.69.2.b83x6t44878u4782

Stiglitz, J. E., Fitoussi, J.-P., & Durand, M. (2018). *Beyond GDP: Measuring what counts for economic and social performance*. OECD Publishing. https://doi.org/10.1787/9789264307292-en

USAID Oceans and Fisheries Partnership. (2019). *Assessing fisheries in a new era: Extended guidance for rapid appraisals of fisheries management systems*. Prepared for the US Agency for International Development by Tetra Tech ARD under Contract No. AID-486-C-15-00001. Retrieved February 9, 2021, from https://www.seafdec-oceanspartnership.org/wp-content/uploads/USAID-Oceans_Assessing-Fisheries_RAFMS-Guide_April-2019.pdf

Voyer, M., Barclay, K., McIlgorm, A., & Mazur, N. (2016). *Social and economic evaluation of NSW coastal professional wild-catch fisheries*. Research project. FRDC 2014/301. Sydney: University of Technology Sydney. Retrieved January 7, 2021, from https://www.uts.edu.au/about/faculty-arts-and-social-sciences/research/fass-research-projects/social-science-fisheries/valuing-coastal-fisheries

Voyer, M., Barclay, K., McIlgorm, A., & Mazur, N. (2017). Using a well-being approach to develop a framework for an integrated socio-economic evaluation of professional fishing. *Fish and Fisheries, 18*(6), 1134–1149. https://doi.org/10.1111/faf.12229

Young, O. R., Webster, D. G., Cox, M. E., Raakjær, J., Blaxekjær, L. Ø., Einarsson, N., Carothers, C., et al. (2018). Moving beyond panaceas in fisheries governance. *Proceedings of the National Academy of Sciences, 115*(37), 9065–9073. https://doi.org/10.1073/pnas.1716545115

INDEX[1]

[1] Note: Page numbers followed by 'n' refer to notes.

Printed by Printforce, the Netherlands